Albert Mata

THE ART OF SUCCESS THROUGH TEAM SYNERGY

Information Technology Management

A guide to managing a Domestic/International systems technology organization

Contents

Preface ix
About the Author xi
Acknowledgements xv

Part I: Information Technology Management
    The Manager Theory 5
    Finding Your Forte 9
    People Skills 13
    Delegating 16
    Ethics and Values 20

Part II: Managing IT Human Resources
    Building your IT department 22
    Cross-cultural Considerations 25
    Individual Employee Development 29
    Managing Behavior 34
    Recognition and Feedback 37
    Retention and Motivation 39

Part III: Creating Team Synergy
    Building Teams..... 43
    Maintaining Productive Teams 44
    Remote Team Synergy 47
    Team Morale 51

Part IV: Technology
    The Goal: 100 Percent Availability 55
    E-Business Strategies 63
    Security 66
    Continuity Planning 74

Part V: Communication and the Business
    Communicating Systems Issues 81
    Communicating Systems Projects and Changes 84
    Communication Formats 88

Part VI: Customer Service
    Aligning with the Business 94
    Dealing with Angry Users 97
    The "No" Circumstance 100
    International Technical Support 103
    Improving Customer Service 106

Part VII: Managing your Environment
    Managing IT Services 108
    Risk Management 111
    Budget Basics 113
    Managing Through Change.. 116
    Inventory Administration 121

Part VIII: Project Management

Take Charge 126
Project Effectiveness 131
Meetings 135
Managing Cross-cultural Projects 137

Part VIIII: Personal Management

Time Management 139

Stress Management 143
Cross-cultural Fitness 145
My Personal Workout 149
Maintaining Your Health.. 151

Appendixes

Recommended Readings 158
Case Study 159
References 169

Preface

The ideas of this book have come from my broad experience in the Information Technology (IT) arena, both domestically and internationally. The techniques and approaches detailed are a compilation of successful practical experiences applied to an ever changing IT environment. In today's technology world, IT managers have to be a leader, a friend, a mentor, an all knowing tech, and at times even a psychologist. These are just a few examples of the many hats the challenging management role requires. There are thousands of IT related books and manuals that can help a manager function more effectively. However, most are related to theories that have become outdated and obsolete due to extreme advancements in technology; these changes create the need for new management techniques. Over the last 15 years there have been dramatic changes in the way a company utilizes technology to do business. Hardware, software, and infrastructure concepts are amongst the areas that have become extremely challenging, even for the most veteran technician. On the other hand, there is one crucial aspect to the systems environment that has rarely changed over many years of business - that is the IT employee. From manual input processes to advanced barcode data entry systems (SQL servers), from *basic* programming to advanced systems (AS/400's), from passive hubs to smart switches and routers (Cisco), the human factor will always play a crucial role in administering a successful information technology environment. Therefore, I have put together many ideas and concepts through years of my work and study that have proven to be effective in building a strong technology team. It is not a summarization of the management process, but rather a compilation of techniques and approaches to many of the most challenging factors facing an IT manager. In addition to this wealth of IT management information and by popular demand, I have also provided a personal guide to maintaining a sound mind and body.

- Albert Mata
El Paso, Texas

P.S. Many readers have reported to me that they use this book as a personalized motivation handbook by simply referencing during stressful systems technology changes and projects. I hope my challenges and accomplishments can help you create a successful and efficient technology team. Good Luck!

About the Author

Albert Mata, MBA/GM, CNA, MCP, PMP, MPM, CIPM, ITILv3, Six Sigma Black Belt is an Information Technology Manager for CSC and having managed technology environments for numerous business accounts. His work experience extends from International Manufacturing Operations to the Business Financial arena. His diverse background, extensive travels, and technical experience make him one of the leading management resources for any Information Systems organization. Also, he is the owner and president of Comtech Technical Support Solutions, Incorporated, an IT outsourcing company specializing in contractual technical support of small to medium scale network infrastructures and voice technology environments.

He is married to the former Elizabeth Delgado and they have one child, Albert III, who is a graduate of San Diego State University and the Music Conservatory of Arts and Sciences. They make their home in El Paso and Dallas, Texas.

Part I: Information Technology Management

> *"No one lives long enough to learn everything they need to learn starting from scratch. To be successful, we absolutely, positively have to find people who have already paid the price to learn the things that we need to learn to achieve our goals."*

The IT Manager Theory

If you are wondering whether you want to be an Information Technology (IT) manager or if you are considering the idea of turning in your technical cap for a leadership hat, pursuing a management path in the technology world is a major career decision. The motivation to make the change from a technical role to a managerial role may come from many directions. Maybe the company has suggested a management position for you. Maybe someone in your life is pushing you to "make more out of your life." Or maybe you are trying to decide whether to get your Masters degree in your technical specialty or go for an MBA instead. Whatever the reason for pursuing a management career, this section will help you decide or confirm whether management is for you.

The Advantages

There are many positives to being a manager in the technology industry. Technology Managers generally are paid more than others are in the organization. They appear to have more power. And the power and pay differences tend to give the position more status or prestige.

- *Pay.* Certainly the top technology manager in a company, the Chief Technology Officer (CTO) is paid more than anyone else in the systems organization. Managers below the CTO are generally paid more than everyone in their group as well, but not always. I managed a group of network engineers in which the very top engineers were paid more than I. Smart companies pay their people based on their value to the company, not on their title or position, and in that company, key engineers were more valuable than their manager.
- *Power.* Most people, including most managers, believe that managers have more power than the people in their groups. While it is true that managers commonly have certain functional authority delegated to them, like setting work schedules for the group, true power cannot be delegated to you from above. You are only as powerful as you are capable

of making your group more successful. And while your ability to <u>lead</u> the group greatly influences it, your power comes from the willingness of the people in your group to grant it to you.

- *Status/Prestige*. In our society, people value titles. A title of Senior Vice President, Enterprise Technology sounds much more impressive than Systems Analyst. However, the technology president may work for a 10-person company and make $90,000 per year while the analyst works for a major hospital supply company, supervises 4 other programmers, and makes well over $100,000 per year.

The Disadvantages

Nobody likes the boss and it is lonely at the top. You are the person who always has to make the decision, right or wrong, and somebody is always out for your job. On top of that there are legal liabilities that non-managers don't have as well as financial restrictions.

- *Lonely at the top.* You are not as close to the employees in your group when you are the boss. You cannot afford to be. A manager needs to be a little removed from the employees in order to objectively make the hard decisions. Many first time managers, promoted from within the group, are amazed at how quickly former friends become cold and distant. Even an experienced manager, brought in from the outside, finds the employees more distant than they are with each other.
- *No immediate reinforcement.* A painter gets almost immediate feedback on whether or not he is doing a good job. Is the paint the right color; is it going where it should. A programmer also finds out pretty quickly whether or not a new sub-routine runs. Management is not that way. Goals are usually more long-term, quarterly or even annual. The real measure of a manager's success comes in the way of long-term successes. Furthermore, an improvement in their people management skill is even more long term and more difficult to manage.
- *Buck stops here.* You may, and in most cases should, have your employees make many of their own decisions. However, ultimately the responsibility for the final decision rests with the manager. When it appeared that insulation might have damaged the space shuttle wing, causing catastrophic failure, it was a manager who had to make the decision. It is the manager's job to make the decision, right or wrong.

- *Somebody always wants your job.* There is always someone after your job. Sometimes several people are. As a first line supervisor, you may have several people in your group who think they could do your job better and are actively working to get that chance. As IT manager of a company, you have several people within your own organization who want your job and more people on the outside who are after it as well. They may not agree with the decisions you made or felt they could have made better decisions. You may have actually made a wrong decision and they will use that as leverage to try and push you aside. The higher you go in any organization, the fewer positions there are at that level and the more competition there is for them.
- *Legal Liabilities.* Managers have legal liabilities that most workers do not. Managers frequently have to sign documents, they have to ensure the workplace is free from harassment; they have to keep their people safe. If a manager fails in any of these responsibilities, they may be held legally liable.
- *Financial Restrictions.* Managers often have financial restrictions placed on them because of their position. The most common of these are the insider trading restrictions. The insider's list at a company is almost exclusively managers. While a worker can exercise stock options or trade in the company stock whenever they wish, the managers on the insider list are restricted to windows of time that exclude immediately before and after quarterly financial results are announced.

If your goal is to be CTO of Cisco, you probably should start now on a management career. If you want to be President of the United States, a management track is not required. Several recent Presidents have managed nothing but their campaigns. If you want to brag to your family about what a success you are, and power, prestige, and money are important to your definition of success, management may be they way to go. If you measure success by friendships and how soundly you sleep at night, a management career can create some challenging issues in that area.

Bottom Line, technology management as a career path is definitely challenging and not right for everyone. You have to like responsibility. You have to enjoy working with people. You have to be able to deal with uncertainty and making decisions when you never seem to have all the facts in time. You must learn to live with being legally or financially responsible for the actions of others. Power, pay, and prestige are great career benefits, but believe me you'll earn it.

Dealing with staff, colleagues, and superiors is not what most people feel they signed up for when they went into IT. However, like it or not, managing people is key to any senior-level position. Choose to do it well and realize that managing an IT department is your opportunity to make a real difference, not only for the people who work directly with you, but also for the entire organization. When it all comes together; when all your people are pulling together toward the same goal and setting new records it can be a great feeling. When you see someone you trained go off on their own and become successful, you can take a certain amount of pride for having helped them get started. Technology management can be frustrating, and a lot of hard work, but it ultimately has its rewards.

Λ

> *"People ask the difference between a leader and a boss. The leader works in the open, and the boss in covert. The leader leads, and the boss drives."*
>
> —*Theodore Roosevelt*

Finding Your Forte

Employees look to management for leadership, direction, and support in many aspects of their position. Therefore, it is very important to keep a handle on both your management and leadership skills. Although there is a wide separation between the two, both are required in order to maintain a stable structure within your team.

The following are some character traits that can assist you in identifying and distinguishing your strengths in both categories:

Personality styles

Managers emphasize rationality and control; are problem-solvers (focusing on goals, resources, organization structures, or people); and often ask questions such as, "what problems have to be solved, and what are the best ways to achieve results so that people will continue to contribute to this organization? They are persistent, tough minded, hardworking, intelligent, analytical, tolerant, and have good will toward others.

Leaders are perceived as brilliant, but sometime lonely; achieve control of themselves before they try to control others; can visualize a purpose and generate value in work; and are imaginative, passionate, nonconforming risk-takers.

Attitudes toward goals

Mangers adopt impersonal, almost passive attitudes towards goals; decide on goals based on necessity instead of desire, and are therefore deeply tied to their organization's culture; and tend to be reactive since they focus on current information.

Leaders tend to be active, since they envision and promote their ideas instead of reacting to current situations; shape ideas instead of responding to them; have a personal orientation towards goals; and provide a vision that alters the way people think about what is desirable, possible, and necessary.

Conceptions of work

Managers view work as an enabling process. They set strategies and make decisions by combining people and ideas; continually coordinate and balance opposing views; are good at researching compromises and mediating conflicts between opposing values and perspectives; act to limit choice; and tolerate practical, mundane work because of a strong survival instinct- they are risk adverse.

Leaders develop new approaches to long-standing problems and open issues to new options. They use their vision to excite people, and only the develop choices that give those images substance. They focus people on shared ideals to raise their expectations, and work from high-risk positions because of a strong dislike for mundane work.

Relationships

Managers prefer working with others, and report that solitary activity makes them anxious. They are collaborative with a low level of emotional involvement in relationships; seek to reconcile differences, search for compromises and establish a balance of power and often relate to people according to the role they play in a sequence of events or in a decision-making process. They basically focus on how things get done. Since they maintain controlled, rational, and equitable structures, others may often view them as inscrutable, detached, and manipulative.

Leaders possess an inner perceptiveness they use to relate to others in an intuitive, empathetic way; focus on what events and decisions mean to participants; attract strong feelings of identity and differences (or love and hate); and create systems where human relations may be turbulent, intense, even disorganized at times.

Self-identity

Managers are basically well adjusted to life. They have an attitude derived from a feeling of being at home and in harmony with their environment, and see themselves as guardians of an existing order of affairs with which they identify and from which they gain rewards. Managers believe their role ties in with their ideals of responsibility and duty, and they are committed to socialization - this prepares them to guide institutions while preserving an existing balance of social relations.

Leaders continually struggle to find some sense of order. They do not take things for granted and are not satisfied with the status quo; their "sense of self" is derived from a feeling of intense separation and is independent of work roles, memberships, or other markers of social identity. They may work within organizations, but never really belong to them.

As an IT manager, there are many hats that you may be required to wear (liaison, mentor, friend, technician, etc.). The key here is to sort out the responsibilities and identify the traits that best describes your personality as a manager or a leader. This is a very important first step in finding the balance necessary to enhance your performance. Furthermore, coming to terms with your character will assist you in effectively applying the information found in the subsequent chapters.

It is my strong opinion that the IT manager's role is best categorized as a leader more so than a manager. I'm sure I may be stirring up some controversy, but the fact remains that any IT management role requires more leadership qualities to obtain a systems related objective. This is due mainly to the need for adhering to the ever change technology environment that currently exist. In other words, you must be the aggressive visionary of technology for your organization. The following further describes the separation.

As a manager:

- *You direct the work rather than perform it.* You were hired to manage the staff's work - not be part of the staff. In the IT world you will occasionally have to roll up your sleeves and work with the team.
- *You have responsibilities for hiring, firing, training, and disciplining employees.* Staff development is an important part of your job. Such development often determines whether staff members stay with an organization or leave for better opportunities. In addition to regular performance appraisals, you should work with each person you manage regularly to determine a career path.
- *You exercise authority over the quality of work and conditions under which it is performed.* As a manager your first obligation is to your people. This means you should ensure that your team is well versed in all aspects of their position in order to perform at their maximum capacity. For example, in the event of a fire, do all your staff members know the exact systems process for disaster recovery? The obligation also means you own your customers, internal or external, the highest quality of outputs.
- *You serve as a liaison between employees and upper management.* This is an example of one of the many hats the

IT manager wears. Among them: coach, mentor, leader, diplomat, and psychologist. In this role, you serve as the link between those doing the work and those who need or benefit from the work being done. The liaison serves as a buffer, a praiser, a translator, and a seeker-of-resources to ensure the work is done more efficiently and the employees are recognized when they complete it.

- *You motivate employees and contribute to a culture of accomplishment.* If you are totally committed as a manger, then you absolutely need to motivate and instill pride in order to create a climate of flourishing innovation.

As a leader:

- *You believe that, working in concert with others, you can make a difference.* Team collaboration is crucial to building a successful systems environment. The successful use of teams is the cornerstone of many IT organizations. The first distinguishing characteristic of a team is its members' full commitment to a common goal and approach. Your overall team objective is to develop team synergy, where team members together achieve more than each individual can.

- *You create something of value that did not exist before.* Being first to create something different, especially in the IT field, demonstrates the courage to challenge the status quo. If you can point to one improvement you have implemented in the last six months, you can rightfully call yourself a leader. Be courageous and daring in a positive and diplomatic way.

- *You exhibit positive energy.* The enthusiasm and charisma you exude will energize others. In addition, the team synergy you have created will be enhanced by the passion you have for accomplishing various tasks. Your staff will gravitate towards that positive energy, which will assist you in meeting you challenging objectives through teamwork.

- *You actualize.* The true leader goes beyond the vision to create a new reality. You must visualize and convey the self-actualization in a manner that makes others believe they can do the same. In the end, the collective actions of the whole team will lead to accomplishing the most challenging of projects.

- *You welcome change.* There is no question that change in the IT arena is inevitable. The best leader is one who not only accepts this concept but also searches further to make a difference. You must find the vacuums and work to fill them. Identify the invisible and lead your group to progress.

Again, in my opinion, the challenges facing IT managers today are best described by the "leader" traits listed above. Most important are the qualities necessary to create team synergy within your organization. In addition to the wide spectrum of traits, all of them admirable and beneficial, there are certain characteristics that all leaders seem to posses. They are:

- *Courage* - Leaders who dare to do something are prepared for opposition.
- *Pride* - Leaders take extensive pride in the atmosphere they create.
- *Sincerity* - Leaders convey sincere concern for other people, genuine interest in subjects other than themselves.
- *Adaptability* - Leaders in the IT world are basically expected to do more with less.
- *Influence* - Leaders know how to influence others, to persuade them to a higher calling.

Above all else, the leader is a person who exemplifies the trait that most management professors regard as the single most important ingredient in the formula of success: "Knowing how to get along with people." Leaders demonstrate adaptability. They can accommodate themselves to different kinds of people, different projects, different places, and different positions.

However, please keep in mind that to lead well you have to manage well. Furthermore, apart from the technical knowledge you need to get the job done, you need other kinds of knowledge if you intend to successfully manage yourself, processes, and the people involved in the processes. You owe it to your followers to acquire this understanding. No matter how much you already know, there is always more to learn. In other words, be a constant hunter. Continuously search for new techniques to help you with your people management skills. You can find approaches worth emulating all around you, not just in the workplace.

△

> *"There are three kinds of people; those that make things happen, those that watch things happen and those who don't know what's happening."*
>
> — *Unknown*

People Skills

Year after year statistics continue to show that the number one reason people quit their previous employer was because of their manager. A painful statistic when you consider how difficult it is to find good people. For a business, this is distressing when you look at the bottom line and unintelligent when managers continue to do nothing about it.

IT managers today walk a thin line and the job is not easy. The responsibilities and demands are more difficult than ever. People expect more; some are plain difficult to work with. Therefore, it goes without saying that those managers that do a good job selecting, training and developing their employees will enjoy higher productivity and lower turnover. The two go hand in hand.

No matter what type of business you are in, soft skills reign supreme and are critical for success. Yet, most managers do a miserable job selecting and training their staff. Additionally, many employee development programs focus entirely on the technical aspects of the job and not people skills. Consequently, some employees remain disruptive and make life miserable for their teammates. Hence, it is your people skills that make the difference in managing your employees.

In addition to employee management, people skills are an important factor in establishing and developing strong relationships with the business. As an IT leader it is a given that you need to be able to build relationships if you want to get positive results with people. It basically takes: highly effective interpersonal and communication skills. And, if you possess what is know as the 'Like Factor' (you are well-liked by others) you will definitely have an edge. But you also need to be sure that you project the right message (i.e. that your words, phrases, and actions reveal admirable qualities and thus, will project a positive image to others).

To establish and build strong relationships, always choose your words carefully. And, as you act and interact, be sure you are projecting the following:

- *Self-assurance.* You need to come-across as completely self-assured if you want to gain support, earn respect from the business, or influence your team. This is a vital quality you must project. Getting others to believe in you begins with believing in yourself. Be sure your self-assurance never comes across as arrogance, but does come across as confidence.

- *Humility.* This is a good companion to self-assurance. While you want to project a feeling of confidence, you also need to project humility and honesty. When you do, you will project that you do not see yourself as being superior to others.

- *Certainty.* You must demonstrate honest belief in your ideas, facts or concepts. Being steadfast builds acceptance. Be sure to refrain from dodging tough questions, or responding evasively when people ask questions regarding important matters. If you are uncertain about something, state that you will find out the facts and get back to them. If there is a decision that must be made, and you need to think about it, state that you will give it thought, and get back to them. Then, when you communicate your decision, project the self-assurance others need to hear from you.

- *Consideration.* Demonstrate consideration for others. (How would this person/people feel if I say/do this?) Projecting consideration will draw people closer to you and help gain their respect. Whenever possible, involve others in your decisions to demonstrate your respect and consideration for their ideas. This is vital to building strong relationships.

- *Integrity.* Without integrity, you will never convince anyone that you are going to do what you say you are going to do. The old adages, walking your talk, practicing what you preach, doing unto others, are all a strong part of projecting integrity. Be sure to resist any temptation to stretch the truth, tell a little white lie, or omit facts that may be inconvenient to mention just to win people over. Refrain from "badmouthing" anyone or you may come across as unprofessional.

- *Inspiration.* If you can inspire people and project a positive attitude, people will gain energy from you and enjoy hearing from you. You do not need to deliver rousing speeches, you simply need to display a positive attitude, be optimistic, and reinforce others to bring out the best in them. Ultimately, you will win them over.

- *Empathy.* Empathy goes beyond compassion. Here is the difference: Compassion is the ability to feel the way others feel. Empathy is having the ability to put yourself in another person's situation, experience their feelings and emotions, and project it to them.
- *Credibility.* Use language that projects your credibility. If you do not, people will doubt you. Keep this in mind too: a 'title' does not necessarily provide you with credibility in the eyes of others. It must be earned.
- *Vigor.* If you project vigor you will be able to release your energy to others, fuel action and endeavor positive reactions from people. Being in the same room with you should provide positive energy to others. You do not have to change your personality to do this. If your nature is 'laid-back,' simply reinforce, appreciate and applaud the thoughts and ideas of others.
- *Trust.* The most important ingredient of any relationship, whether it is business or personal, is a shared sense of trust. You will never be able to establish or develop any relationship without it, for trust is the foundation for reliability, dependability, and good faith.
- *Honesty.* This is critical to developing a relationship. When things go wrong or a problem occurs, honesty is always the best policy. In fact, if a problem occurs and you solve it the right way, it can make the relationship stronger. If you make a mistake, own up to it, or you will lose face. Apologize sincerely and hopefully the person will be understanding and accepting. Without honesty, there can be no trust.

Remember, a GREAT leader is able to influence others to go WITH them. Therefore, on a day-to-day basis, you must strive to gain respect from each employee if you want to be able to influence your team to follow your lead. Furthermore, by improving your people skills, you can build better relationships within and outside your organization, (and in your personal life too!). Maintain a constant awareness of your personal interaction, ensuring that your communication of your words, phrases and actions are projecting the right message.

Λ

> *"The man who knows how, will always have a job. The man who also knows why, will always be his boss."*
>
> *— Ralph Waldo Emerson*

Delegating

The classic definition of a manager is one who gets things done through other people. A manger's regular routine usually consists of planning, directing, controlling, hiring, delegating, assigning, organizing, motivating, and disciplining. These are just a few examples of the many responsibilities required on a daily basis. No matter what area of business you may manage, the focus is basically to ensure and help others get their work done.

However, the information technology manager is definitely one of the most challenging positions of any management role. One aspect that makes it elite from other management roles is the people. IT managers deal with one of the most complex and demanding personalities in all careers, the mind of a programmer, engineer, or technician (a.k.a. the Techy). Systems technology people are basically analytical by nature. They break down and analyze everything that comes their way. Thus, creating challenges for effective delegation.

Many IT managers find delegating an IT staff a challenging experience. Instead of just being responsible for one's technical work, now one is responsible for the work done by others. In addition, there is a strong tendency to have little faith in others' abilities and sometimes less faith in oneself to manage others and get things done, especially during high profile technical projects. Many times we retreat back to the familiar and try to do all things ourselves.

Early in my management career I myself found it very difficult to step away from the technical aspects of a position and focus on the administrative tasks involved with managing a team. While this may work or appear to work in the short run, in the long term it is a sure path to disaster. Being able to successfully delegate work and ensure that it is done with a minimum of fuss is crucial for any management role. Furthermore, IT managers are not measured based on their individual contributions but on the contributions of their team. So, even if you put in the longest hours, if your team is proceeding nowhere you are in trouble! On the other hand, a successful manager will be able to motivate their teams and help them successfully complete tasks without too much heartburn and overtime.

Delegating with confidence

How does one go about being successful in delegating work? In my experience both as a new manager and as an experienced mentor of other managers, this can be one of the hardest things to learn and do. Not only do you need to learn to hand over control, you also have to ensure that the team is ready for these responsibilities. It is effectively a double challenge.

Delegation always involves some risk, which means you have to live with the consequences of someone else's decisions. Take the following steps to increase your comfort level and improve the person's chances for successful performance:

- *Clarify what you want to delegate.* Describe in unambiguous terms the work you want the other person to perform; also explain what you do not want the person to do.
- *Choose the right person.* Determine the skills and knowledge you feel a person must have to perform the task successfully, and don't delegate the task to a person who lacks these skills and knowledge.
- *Make the delegation correctly.* Explain the work to be done, how much effort you expect the person to expend, and the date by which the work is to be completed.
- *Monitor performance.* Set up frequent, well-defined checkpoints at which you can monitor performance; monitor according to that schedule.

Delegation does not have to be an all-or-nothing proposition where you either make the decision yourself or withdraw from the situation entirely. Consider the following six degrees of delegation, each of which builds on and extends the ones that come before it:

- *Get in the know:* Get the facts and bring them to me for further action.
- *Show me the way to go:* Develop alternative actions to take based on the facts you have found.
- *Go when I say so:* Be prepared to take one or more of the actions you have proposed, but do not do anything until I say so.
- *Go unless I say no:* Tell me what you propose to do and when; take your recommended actions unless I tell you otherwise.

- *How did it go?* Analyze the situation, develop a course of action, take action, and let me know the results.
- *Just go!* Here is a situation; deal with it. I don't want to hear about it again.

Each level entails some degree of independent authority. If the project leader asks you to find the facts about a situation, you choose what information sources to consult, which information to share with her, and which to discard. The primary difference between the levels is the degree of checking before taking action.

Team Delegation
Projects, tacks, and issues come in many sizes and forms. Many times you will be required to delegate large portions of a situation to a team as a whole rather than by individual task. Although it may sound easier than breaking it down by task and team member, the risk factors for losing control of the situation grows many times over. The following are seven steps that can help you meet the Challenge:

1. *Identify team goals:* Meet with your director to understand what you will be responsible for and how you and your team will be measured. One way to get a good handle on this is to write it down. List the important goals and measurements to be used and get agreement from them. This will make it easier later on when you have to do performance reviews as well as decide and possibly fight for bonuses and increases for yourself and your team.

2. *Identify team strengths and weaknesses:* Meet individually with each team member to get an idea of their strengths and weaknesses. Also, get an idea of the management styles they work well with.

3. *Build enthusiasm and team spirit:* Once you have a good idea of what is needed, get the teams together for a meeting. Ideally, you should try to do this offsite where people are less likely to be interrupted. The idea is to get everyone on the same page and ensure that each member in the team understands what his/her goals and responsibilities are. Ask people for ideas on how things could be made to work; decide problem escalation paths and status reporting. The aim is to get buy-in and ensure that everyone agrees with how things should be done.
   One of your major responsibilities will be to ensure that these rules are followed and that they work. Also, get

agreement on contingency plans if things do not happen as expected. You may need to change certain agreements and relationships based on how the team performs. Remember that, as far as senior management is concerned, YOU ARE responsible for making things work. Make sure that the team understands this—they will be more willing to support you when you need to make the tough decisions. On a regular basis, try to get the team to do something fun together, especially after meeting major goals or milestones. You can juice up the competitive spirit by giving recognition and small gifts to those who did something exceptional. You may want to get the team to vote on those they think did exceptional things and kept up the team spirit.

4. *Identify your replacement:* This may sound odd, but you need to find a confidant. The idea is to find someone, ideally within the team, who can take charge when you are out sick, on vacation, or business travel. Based on people's participation in the team, their experience, willingness, and so on, you should be able to identify good candidates within the team. In my experience, asking someone not on the team to be your replacement may not work well. They will not be so familiar with team dynamics, the history, the people and so on. Plus, they may have other responsibilities that will take up most of their time. Unfortunately, on occasion you may experience some jealousy on the team because you are spending more individual time bringing your replacement up to speed on their responsibilities. However, during your absence, your replacement will play a crucial role in maintaining team standards and business liaison communication. Therefore, it is imperative that your replacement obtains a mirror image of your expectations for managing the department.

5. *Train the team:* Start grooming those identified to act as your replacements to take on more responsibility. Some ways are to create sub-teams that they are responsible for. Also, try to make every team member take on more responsibility and control over his or her work. This will build their confidence as well as make them enjoy their work more. Some people will adapt quickly, while on the other extreme, it can be very scary to people who are more used to top-down management that clearly defines what they can and cannot do. However, when working on

technical projects, and especially when using agile methods, it is very important that each team member is very self-directed and responsible for their work. After all, people are not producing widgets on an assembly line!

6. *Be very proactive in finding and handling problems:* Try to have regular meetings with the team on a daily or at least weekly basis. However, keep them as short as possible because people resent taking time off from doing their work, especially when deadlines loom. Keep an eye out for conflicts within and between teams. Act fast to nip them in the bud. Sometimes, these are misunderstandings that can be quickly cleared. At other times, you may need to move people around if two of them have opposing work styles and find it hard to work with each other.

7. *Plan ahead:* One thing to watch out for is changes or other situations that can affect the team. Examples could be staff cutbacks, need for additional staff, new projects, project cancellations, and so on. Try to get in the loop to find out what may be coming down the line. Try to meet with folks in other groups and departments and understand more about their jobs and challenges. This will give you an idea of what is important for the organization as well as help you identify new opportunities where you and your team could help.

For many IT managers, it can be quite a challenge to figure out what they need to do to keep things efficiently moving along. One of the toughest obstacles to overcome will be yourself—you have to trust other people to do the work. Another challenge will be to effectively verify that the work is done while providing support and encouragement to your team. The seven-step process outlined above should help you enhance your delegation skills and become more effective as a team leader.

Λ

> *"Integrity without knowledge is weak and useless, and knowledge without integrity is dangerous and dreadful."*
>
> — *Samuel Johnson*

Ethics and Values

Ethics in the field of technology is concerned with questions of value (i.e., judgments about what human behavior is "good" or "bad" in any given situation). Ethics are the standards, values, morals, principles, etc., which are used to base one's decisions or actions on; often there is no clear "right" or "wrong" answer.

There needs to be awareness among technology professionals as well as the general user community of the importance of ethical practice and social responsibility. IT personnel, including managers, who aspire to become successful professionals "need to understand the basic cultural, social, legal, and ethical issues inherent in the discipline of computing. You should understand where the discipline has been, where it is, and where it is heading. You should understand their individual roles in this process, as well as appreciate the philosophical questions, technical problems, and aesthetic values that play an important part in the development of the discipline. Furthermore, your team must be able to anticipate the impact of introducing a given application or device into a given environment. Will that device enhance or degrade the quality of life for the customer (user)? What will the impact be upon individuals, groups, and the organization(s)?"

Unfortunately, many IT professionals view computer technology merely as a personal intellectual challenge, a test of their ability to solve logical problems and to control the technology. Such a narrow approach to computing implicitly demonstrates the lack of ethical responsibility or social obligation in the practice of computing skills. It is important to become aware of the tremendous responsibility to other people that comes with the practice of your expertise and learn how to make these critical judgments. To accomplish this, it is necessary to make a strong case for ethical behavior in the context of the system technology profession and to provide activities and exercises to enable your team to develop skills in ethical and social analysis.

One of the most effective ways to introduce ethics to your team is by the use of a variety of ethics activities in technical courses taught by local colleges. This demonstrates to your IT staff that ethical/professional/social content is a fundamental component of the ethics and values discipline. This also helps to dispel the misconception that ethical, professional, and social issues are less important or separable from computer science and information technology.

Literature obtained for training should relate to information technology as the basis of ethical practice. Using information technology to present social and ethical issues makes IT staffs more interested in the issues and allows for more variety in the presentation of them. Hence, helping them realize just how connected technology is to social and ethical issues.

I do not recall where I located this map, but I have found it helpful in recognizing and referencing the many areas of concern related to ethics and technology.

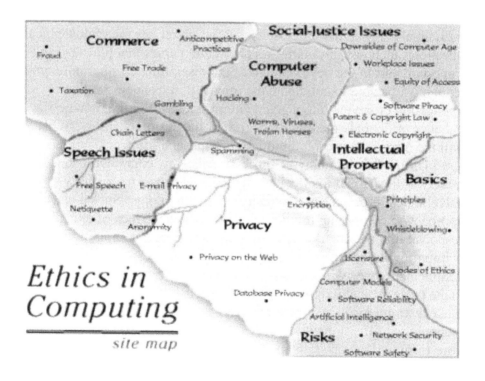

IT Management Lessons

- Management theories require a commitment. If you choose to be an IT leader keep an open mind about what is required in your environment. Manage your employees, but lead your team.
- Effective delegation is quickly becoming a lost art. IT leaders, especially those with large engineering groups, need to master and maintain those delegation skills.
- Ethics, Ethics, Ethics, today more than ever IT leaders need to have full control of what occurs in his/her technical environment. Know what your employees are doing, but more importantly, know what the business is doing.

Part II: Managing IT Human Resources

> *"A few honest men are better than numbers."*
>
> *-Oliver Cromwell*

Building your IT Department

As you begin to interview applicants to fill a job position in your firm, you need to be sure what you are looking for in these applicants. If you are like many technology managers, your concept revolves around terms such as hardworking, persistent, confident and dependable. Most managers feel they cannot go wrong trying to hire people with traits such as these. The truth of the matter is, you can. The problem is that traits are not necessarily good predictors of future job performance.

There is no <u>perfect </u>formula for finding the right person for the right job. However, there are many ideas, rules, and suggestions to help screen out the most productive or qualified candidate.

Rule number one is clear and supremely counterintuitive: *don't ever, ever hire somebody just like yourself.* Why is that so hard? Because from the beginning of technology as a business resource, IT mangers have been unconsciously cloning themselves, creating an environment of techies from impressive schools rather than impressive personal and technical background experiences. Managers and companies that do that often create a vanilla like environment with little potential for intuitive variety.

The following are nine other rules to keep in mind:

> *Hire for Attitude Rather than Skill.* Teaching skills is a snap compared with doing attitude transplants. Among the qualities you'll want most is a fierce sense of optimism.
>
> *Look for Renegades.* In interviews, ask when the person has been in trouble. The obedient employee will be of limited use to you in this change-up environment.
>
> *Hold out for Results.* Never hire someone with good potential but questionable habits, thinking you can change him or her. As in choosing mates, what you see now is what you get forever.

Go for a Sense of Humor. The potential hire that cannot laugh easily, particularly at him or herself is going to be a very dull and probably rigid employee.

Fill in the Blanks. Look carefully at the aggregate strengths and skill gaps of your teams in various work units, and go for the qualities and styles that are missing.

Test Drive. Do not be satisfied with references. Remember that many of the most glowing references are given for people others are eager to dump. Include day-long simulations as part of your interview process, or invite applicants to provide you with a portfolio of their best work.

Stock the Bullpen. Keep an eye out for prospects before the need arises. Do not wait until a vacancy occurs. Keep a pool of potential employees under the watchful eye of somebody who is responsible for hiring. Evaluate your recruiting team in terms of how well they keep the bullpen ready. And tell them never to turn away an interesting candidate with the line, "We do not have any positions open right now."

Push Harder for Diversity. Make certain you are spreading your net wide enough to find those high-potential, but different, candidates who generally do not work in regions close to you. Ask your HR group what contacts and periodicals they are using to interest potential hires. "We don't know where to find people different from us" is a costly excuse.

Listen. Most interviewers talk way too much. When a candidate finally gets to you, listen for the "story line" of his or her life, at home and at work. It has been said that being a leader is like practicing psychiatry without a license. I believe the concept holds just as true during the hire stage.

Sorting the candidates

Recruiting for an employee to fill a technology position can be a frustrating and time consuming process. From writing the ad, which has to be exciting and challenging, to the actual interviews with likely candidates, it is guaranteed your time will be stretched to the maximum for a period of time.

Here are 5 more basic tips to help you hire a great employee:

1. When you review a resume, look for length of time on the job: a candidate with several short- term employers (less than a year) could mean a lack of commitment on their part. Of course, with all the recent layoffs, it could just mean they got caught in the fallout. Nevertheless, it is not necessarily a given that they were laid off because the company was cutting back. Star technical performers are not laid off if the company can avoid it. Be sure to check those references carefully, especially if the company is still in business. You don't want someone else's "deadwood".

2. Look for gaps in employment and ask for an explanation. Long periods out of work could signal a time out to update their education - or it may indicate some criminal background.

3. Watch the employment dates for "overlaps". This could be a simple error, but also might indicate that the candidate is not being truthful about previous employers. Ask the person to explain it, and be sure to call those employers and verify dates of employment with them.

4. Have a copy of the job description at hand when you review the resumes. The more skills that match your job description, the more likely you will be to have a great match. Focus on what they have done in the past that matches what you want them to do. If you need seasoned technicians who have developed E-Commerce web sites, watch for that on their resume. Write a list of things you want them to do and then ask questions that will get those answers (or the lack thereof).

5. Sort your likely resumes into two stacks, one for those that look perfect to you and the other for those that look good. Call the perfect ones on the phone and ask them why they want to work for your company. Pay attention to your instincts - if you need a Help Desk Administrator, it is important that the person has a pleasant phone voice, that they are enthusiastic and articulate. If they do not make you feel good when you are speaking to them on the phone, they will not make a very good first impression on the customers who call for support.

While these tips are not the whole picture in hiring a great employee, using them increases your chances that you will find the perfect match for your department or company. With the present restrictions on information that a previous employer is allowed to give you, it is important to pay attention to all the details you get from the candidate. It will make the difference between having a great employee working for you or just having a good one.

‣

> *"Arrogant people are fearful people. The more arrogant, the more fearful. In truth, arrogant people are fragile.*
>
> *--Dr. Theodore Rubin*

Cross-Cultural Considerations

Culture in general is concerned with beliefs and values on the basis of which people interpret experiences and behave, individually and in groups. Broadly and simply put, "culture" refers to a group or community with which you share common experiences that shape the way you understand the world. Thus, the same person can belong to several different cultures depending on his or her birthplace; nationality; ethnicity; family status; gender; age; language; education; physical condition; sexual orientation; religion; profession; place of work and its corporate culture. Furthermore, culture is the "lens" through which we view the world. It is central to what you see, how you make sense of what you see, and how you express yourself.

The more you understand human nature - "what makes people tick" - the more you can bring out the best in your workforce. This is especially true in diverse work environments. As the US becomes more diverse and companies employ more international staff, the issue of communicating and interacting effectively with people from other cultures is becoming significantly important to the success of an organization.

Understanding and effectively interacting within the cultural groups to which we belong is like walking through a minefield. Being culturally savvy means much more than just understanding the culture of other groups or countries. It means understanding who you are and your own cultural dynamic.

Consider the following "culture" qualities, and the impact on your life:

- Where you were born
- Your nationality & heritage
- How you were raised & your family life
- The schools you attended
- Your religious preferences
- Your profession
- Your company and its corporate culture
- Your gender

Having respect for cultural differences and learning basic characteristics of other cultures can help you avoid misunderstandings and unintentionally offending others. Hiring, as well as inheriting, a diverse staff can be a challenging aspect of any management position. It can totally change your employee strategies and long-term development goals. Although there is no overnight fix, the most important thing you can do for your team members is to gain a solid understanding of their cultural differences and backgrounds.

Make time to:

- *Learn the language* - try to learn a few common phrases in the languages of your international staff. This shows you are interested and helps them to feel more comfortable. Language differences can make communication difficult, but if you are creative you can find ways to communicate effectively. Talk slowly and clearly. It may sound simplistic, but do not shout at people if they do not understand what you are saying. Restate the statement in a different way or repeat it more slowly. Use other methods of communication, such as making drawings, demonstrating or acting out, or using hand motions.

- *Learn the culture* - People in the U.S. generally call others by their first names, but this is not true of all cultures. Ask staff members how they would like to be addressed. Make sure to call them by their real name, not the U.S. equivalent or nickname unless they prefer it. If you have difficulty saying their name, ask for the correct pronunciation. Many cultures have difficulty saying "no" to a request, and some internationals may say "yes" when their answer is really "no." Carefully phrase questions so that they can be answered positively. For example, say "What can I do to make your work environment more enjoyable?" instead of asking "Do you like working here?"

- *Learn communication techniques* - communication is an area that can be especially challenging for those uninformed about cultural differences. A simple nod of the head or smile may be interpreted as something you had not intended. For example, around the world a smile can relay many emotions, not just happiness or pleasure as in the U.S. In Japan, people smile when they are sad, angry, confused, and happy. Asians smile to show disagreement, anger, confusion, and frustration. Some people from Japan and Asia will not smile for official photos, such as passport photos, because these are considered serious occasions and they do not want to look as though they are not taking the situation seriously.

- *Explain your expectations in regard to time and punctuality.* Cultural background influences what people consider being on time, late, and early. Make sure that staff members know when you expect them to be at their job.

Learning about other cultures and having respect for individual differences can go a long way in creating a successful manger/employee relationship. Be objective, have an open mind, and your organization will benefit from an international influence.

Culture and Trust

Research indicates that there is a strong correlation between components of trust (such as communication effectiveness, conflict management, and rapport) and productivity. Cultural differences play a key role in the creation of trust, since trust is built in different ways, and means different things in different cultures.

For instance, in the U.S., trust is "demonstrated performance over time". Here you can gain the trust of your colleagues by "coming through" and delivering on time on your commitments. In many other parts of the world, including many Arab, Asian and Latin American countries, building relationships is a pre-requisite for professional interactions. Building trust in these countries often involves lengthy discussions on non-professional topics and shared meals in restaurants. Work-related discussions start only once your counterpart has become comfortable with you as a person.

Cultural differences in multicultural teams can create misunderstandings between team members before they have had a chance to establish any credibility with each other. Thus, building trust is a critical step in creation and development of such teams. As a manager of a multicultural team, you need to recognize that building trust between different people is a complex process, since each culture has its own way of building trust and its own interpretation of what trust is.

Culture and Rapport

Rapport is essentially meeting individuals in their model of the world. It starts with acceptance of the other person's point of view, their state and their style of communication. Rapport is the ability to be on the same wavelength and to connect mentally and emotionally. It is the ability to join people where they are in order to build a climate of trust and respect. Having rapport does not mean that you have to agree, but that you understand where the other person or people are coming from.

Rapport is the first step in good communication. The key to successful rapport is not only to be a creative communicator, but also a creative receiver. A sender cannot decide what the signal will actually mean to the other person, only what they would like it to mean. When you mean one thing and the other person shows you by their response that they have a different meaning, nobody is wrong and nobody is to blame.

Rapport is also the key to influence. To influence you have to be able to appreciate and understand the other person's standpoint. And these work both ways: I cannot influence you without being open to influence myself. Rapport works best when it is a philosophy - a way of dealing with people and a way of doing business at all times - in contrast to doing rapport as a technique in a support meeting or when there is a problem. Having rapport as a foundation for the relationship means that when there are issues to discuss, you already have a culture in place that makes it easier to talk them through and thus to prevent issues from developing into complaints, objections or problems.

Global Communication

Frequent business trips across the border and/or abroad are now a constant norm in today's growing global management environment. The main increase in travel is due to companies moving their business to regions where face-to-face meetings are preferred over other communication methods. For example, Mexico considers phone conferencing and e-mail to be an impersonal way of discussing important business issues, especially if it is with a US partner. Facial expressions and body language play a major role in determining the direction of a conversation. Thus, making it significantly important for an IT manager to be well versed in the cultural communication methods of his/her business associates.

Although more and more people around the world are learning English and becoming more proficient communicators, serious business meetings and negotiations are still often conducted with the assistance of interpreters. Never assume that people do not speak English (or any other language), just because they do not say anything. Making this assumption can lead to embarrassing situations. In Japan, for example, there is often a spokesperson for meetings.

On the other hand, do not assume that everyone speaks English (or any other language), even if you find people nodding and acknowledging your comments. This may just be their way of being polite. Many people, even though they may be quite fluent in a foreign language, still prefer to use an interpreter for several reasons:

- Even a few mistakes can be costly or embarrassing. In Asian cultures, there is always the risk of loosing face!
- Using an interpreter is a way of buying time, giving you the opportunity to formulate a response or evaluate what is being said.
- There is a distinct advantage to communicating in your own language, unless you are totally bilingual.

Be sure your interpreter is well versed in your business and understands the goals and objectives of the meeting or presentation. (And) always hire the best available, your business and reputation as a manager depends on it. Other techniques to keep in mind with an interpreter include:

- Pause after about 60 seconds, after a thought is complete (if not too long) or after major points have been made.
- Watch your interpreter to be sure he/she is keeping up with you.
- Avoid long, complex sentences, and do not use slang, jargon, or colloquial expressions.
- Avoid jokes and humorous stories. Humor does not travel well.
- Advise your group that only one person should speak at a time.

If you have a good understanding of the other language, you may be able to follow along and spot any errors in the translation. However, be cautious of embarrassing your interpreter. Maintain eye contact and address remarks to your audience or counterparts, NOT the interpreter.

Bottom line, good interpreters can be very helpful with cross-cultural communication. For example, Americans are much more direct and casual than the Japanese. Your interpreter will be able to put a cultural filter on the communication. Understanding the cultures in which you do business is definitely a plus and very helpful when working over the border or abroad.

⚠

"A lot of people have gone farther than they thought they could because someone else thought they could."

--Zig Ziglar

Individual Employee Development

Development is the process of enhancing personal and professional skills and experience in a way that increases the employee's value to the organization. Employee development is achieved through one or more of the following: academic and professional development, work assignments (short and long term), and various forms of self-study.

In training and developing employees, it is important to assess needs and goals and to plan developmental strategies. These are the two steps that can make development an individualized, systematic process rather than the haphazard process that usually evolves. There are many tools and techniques available to develop a specific plan of action for learning and career development. Two of the most popular are Individual Development Plan (IDP) and Personal Performance Review (PPR). An IDP is a document used to track an employee's progress toward career, professional, and personal growth. It is a tool for recording development goals, plans for taking advantage of developmental opportunities, and the outcomes of investing in those activities.

A well-prepared IDP might include occupational exploration and research, professional development, skills training, and formal education. Consider the following guidelines for creating and implementing an IDP:

Know the Purpose of the IDP

Prioritize and develop a plan of action to reach short- and long-term goals.
Focus personal efforts in the areas that have been selected.
Identify, outline and use resources.
Create an action plan that is clear and achievable.

Put the IDP into Action

Discuss the IDP with your employee and other appropriate individuals such as a human resource counselor. Use this time to state goals, clarify expectations and agree upon modifications as needed.

Work the plan – take a first step.
Evaluate plan along the way and modify it as needed.
Expect obstacles and assist the employee to overcome them.
Celebrate the successes along the way!

As an IT manager you are responsible for identifying the professional development goals and opportunities, provide guidance and advice in planning, and retaining a copy of the IDP to track employee progress. Example 2-1 is a generic IDP format example to help you with this process. Remember, IDPs are living documents - they can be revised and updated throughout the year as priorities and goals change, just as you might change your route on a roadmap.

Example 2-1

Individual Development Plan (IDP)

Name _____ Position _____ D

Department _____ Manager _____

Goals What will be achieved?	Competencies What will be learned?	Actions What steps will be taken?	Resources What is needed (money, time, etc.)?	Time Frame	
				Start	Complete
Short-range Critical within present position (1 year)					
Mid-range Important for growth within present or future position (2 years)					

Long-range Helpful for achieving career goals (3 years)					

IT managers are also responsible for assisting employees through career and performance development discussions, usually as part of an annual and/or midyear personal performance review process; assisting employees in identifying opportunities consistent with the individual's and the company's objectives. Thus, a more formal review process is necessary for identifying and addressing critical performance factors. Figure 2-2 is a basic example of a PPR.

Example 2-2

Performance Review and Development Plan

Results of Business Practices for Performance Period

Employee's Name	Performance Period xx/xx/xx-xx/xx/xx
Action oriented: Produces results, overcomes obstacles and works effectively, appropriately escalates critical issues to management	
Comments:	
Adaptability: Can effectively cope with change, and demonstrates ability to adapt to new strategies, plans, programs, cultures and philosophies.	
Comments:	
Communication Skills: Develops clear, concise and grammatically correct written and verbal communication.	
Comments:	
Problem Solving: Determines problems from symptoms in a logical and methodical manner; provides alternatives.	

Comments:
Planning and Organizing: Anticipates and overcomes problems; coordinates and defines plans supporting strategic goals.
Comments:
Team work: Constructive business relationships; influence and help others.
Comments:
Customer service: Shows concern, maintains honesty and integrity, responds to customer needs.
Comments:
Job knowledge: Exhibits job comprehension, expertise and skills, maintains technical currency in field.
Comments:
Leadership: Is decisive, influential and recognizes others.
Comments:
Strategic Planning: Develops strategies, establishes objectives and meets priorities.
Comments:
Decision making: Analyzes problems and takes appropriate action.
Comments:

Training Plan

Employee's Name		Performance Period xx/xx/xx - xx/xx/xx
DEVELOPMENT PLAN		
Completion Date	*Development Activity or Training Course*	**Responsible Person**

Skills Assessment completed: *(enter date and attach assessment)* _____ *(assessment date)* _____

Goals for Upcoming Performance Period

Employee's Name	Performance Period xx/xx/xx - xx/xx/xx
Performance goals and targeted completion date (list in priority order)	
Goal #1: **[Describe expected result]**	
Results at year end:	
Goal #2: **[Describe expected result]**	
Results at year end:	
Goal #3: **[Describe expected result]**	
Results at year end:	

Creating a development plan from the PPR feedback serves as a commitment to work toward specific goals, a mechanism to communicate goals, and a means to monitor progress in achieving them. Employee goals will be more focused, training and development opportunities more aligned, and partnership with your employee enhanced.

The PPR process involves more than just completing a form. Generally, PPRs are useful for documenting career and performance planning discussions between you and your employee. During the performance review and planning session, the PPR will be documented and signed by both you and your employee. Periodically during the performance year, you and your employee should review the PPR on a one-on-one basis to assess your progress and address any positive and/or negative patterns.

Encouraging outstanding performance from employees does not happen by coincidence. It is the result of managers who take the time with their employees to craft performance expectations and who provide those employees with ongoing feedback and opportunities to develop.

⚠

> *"Blessed are the people whose leaders can look destiny in the eye without flinching, but also without attempting to play God."*
>
> *--Henry Kissinger*

Managing Behavior

Most of us have a strong belief in the power of traits to predict behavior. We know that people behave differently in different situations, but we tend to classify people by their traits, impose judgments about those traits (e.g. being self assured is "good", being submissive is "bad"). (And) we make evaluations about people based on these trait classifications.

IT Managers often do this when they make hiring decisions or evaluate current employees. After all, if managers truly believed that situations determined behavior, they would hire people almost at random and structure the situation to fit the employee's strengths. There are two problems with using traits in managing the behavior process. First, technology and business organizational settings are strong situations that have a large impact on employee behavior. Second, IT individuals are highly adaptive and analytical. Their personality traits change in response to organizational situations.

Business psychologists and executive coaches have studied and found tested, effective ways to help people recognize and correct their bad habits. These are destructive behavior patterns and refer to deeply rooted psychological flaws. For instance, an employee whose psychological makeup translates into consistently problematic behavior limits his or her success and possibly destroys his career. The researchers identify the root causes underlying these behaviors as a mix of an individual's genes and environmental influences. Four psychological processes include an inability to understand the world from the perspective of other people, a failure to recognize when and how to use power, a failure to come to terms with authority, and a negative self-image.

I'm sure you can associate some of these behavior patterns to one or more of your employees. If so, you should understand that your work in helping an employee change a negative behavior pattern won't happen overnight, but your efforts will help a valuable employee become the most effective. Managers can do a great deal to harness the power and energy of a troublesome employee, while working with them to minimize the negative effects of the social style. Below are some suggestions to determine whether there is a lot of good, a lot of bad or a lot of ugly.

- *Get your signals straight.* You may not be able to harness the person, but you can guide them. The trick is to be clear about how you see the person contributing to the organization, recognizing that his contribution may be different from other employees. The employee needs to know what you expect, and what you need.
- *Work for respect, not authority.* Your formal authority may not have much impact on a negative attitude. Don't expect the employee to respond to your requests simply on the basis of you being the boss. What will have an effect is developing rapport and mutual respect. This means dialogue, and a willingness to listen to what the employee has to say. It means asking many questions. It also means showing that you value their contributions.
- *Give feedback.* A troublesome employee doesn't usually intend to be obstructive. They appear so because they simply don't think about how they may be affecting those around them. For this reason, it is important that they receive feedback that will focus them on how they are doing. If an employee is obnoxious in a meeting, they need to be told. The best way to communicate feedback is to talk about basic principles and values and then move to specifics. Feedback isn't just about negative behavior, it is about letting the person know that his or her contributions are appreciated and valued. If you want to keep an employee contributing positively, you need to let him know.
- *Dealing with the ugly.* If an employee is unskilled, not very competent and obnoxious, you have a performance problem that must be addressed. It this person is allowed to run roughshod over everyone without contributing anything positive, the entire organization can be poisoned. There will be situations where the best course of action is to encourage the person to move on, particularly if they are constantly disruptive.
- *Champion and protect.* Remember that the individualist or difficult employee tends not to belong to any particular group, and so doesn't receive a lot of group support. He relies on the strength of his ideas rather than social support. If you value the positive contributions of your employee, you will need to point out these contributions to more conventional employees, particularly in group situations and meetings. Show that you

value the ideas and creativity, even if you don't like the way
the comments or ideas are presented.

- *Set limits (Or Try).* The employee is going to need reminding
  that there ARE organizational goals that are important. Help
  the employee focus on these goals as important, relevant and
  valuable. Don't appear arbitrary, but appeal to principles and
  values they may have.

The individualistic or negative employee can contribute positive
and negative things to an organization, and can be a blessing or a curse
to any IT manager. Much of what determines what you will get is how
you as a manager tolerate the quirks of the employee. If you can
harness the energy and commitment of a negative employee, they can
play an important role in helping the organization shake off the inertia.

Here are some ideas for interacting with people you perceive as
difficult. They may be helpful in managing conflict situations in your
IT work place.

- *Assess the situation.* Ask yourself the following questions:
  Are you dealing with a difficult person or with a situation
  that is temporarily bringing out the worst in an ordinarily
  non-difficult person. Have they acted the same in three
  similar situations? Is this person reacting in line with
  his/her cultural influences? Are you reacting out of
  proportion to what the situation warrants? What particular
  incident triggered the difficult behavior? Will direct, open
  discussion relieve the situation? By asking yourself the
  above questions, you can identify if you are actually
  involved in dealing with a difficult person or situation.
- *Stop wishing they were different.* We all tend to think that
  other people are basically like ourselves, that they have
  similar values, assumptions and feelings. When they do not
  act as we expect, or would like them to, we may tend to
  believe that their behavior is the result of hostile intentions
  or a faulty personality. Thus, we think they should change.
  When they don't change, we become frustrated with them.
  What we need to do is give up the magical wish that the
  other person were different and choose behaviors for
  ourselves, which will likely gain us a better, more
  productive relationship.

- *Put some distance between you and the difficult behavior.* The goal is to obtain a detached and distanced view of that difficult person, while they are being difficult. Picture yourself looking at the person through the wrong end of a telescope, placing them far away. Having a distanced perspective does not mean being unfeeling or non-understanding. A distanced perspective helps you see their behavior as separate from yourself. In order to achieve a broader perspective on a difficult person; try to see things from their view. Imagine how life looks to that person and relate that perspective to comparable experiences in your own life.

- *Formulate a plan to interrupt the interaction.* After you have developed some distance and some understanding of the other person's behavior, prepare a strategy for dealing with it more effectively. Assess if there is a negative interaction cycle operating. A negative interaction cycle is one where a negative encounter between two individuals, or an individual and a group, spirals into an increasingly negative and unproductive cycle of interaction. You can also develop a positive interaction cycle. The major leverage you have for coping with the difficult behavior of others is your ability to change the nature of the interaction. You can basically do this by changing our behavior towards the other individual.

⚠

> *"The deepest principle of human nature is the craving to be appreciated."*
>
> *--William James, American Psychologist*

Recognition and Feedback

Employee recognition is not just a nice thing to do for people. Employee recognition is a communication tool that reinforces and rewards the most important outcomes people create for your business. When you recognize people effectively, you reinforce, with your chosen means of recognition, the actions and behaviors you most want to see people repeat. An effective employee recognition system is simple, immediate, and powerfully reinforcing.

Recognition is about feeling appreciated. It is knowing that what you do is seen and noted, and preferably by the whole team as well as by you, the manager. In opposite terms, if people do something well and then feel it is ignored - they will not bother to do it so well next time (because "no one cares").

When you consider employee recognition processes, you need to develop recognition that is equally powerful for both the organization and the employee. Five of the most important guidelines for effective recognition include:

- Establish criteria for what performance or contribution constitutes rewardable behavior or actions.
- All employees must be eligible for the recognition.
- The recognition must supply the employee with specific information about what behaviors or actions are being rewarded and recognized.
- Anyone who then performs at the level or standard stated in the criteria receives the reward.
- The recognition should occur as close to the performance of the actions as possible, so the recognition reinforces the behavior you want to encourage.

The feedback you give your team about their work is fundamental to their motivation. They should know what they do well (be positive), what needs improving (be constructive) and what is expected of them in the future (something to aim at). And while this is common sense, ask yourself how many on your team know these things, right now?

Perhaps more importantly, for which of your team could you write these down now (try it)?

Your staffs need to know where they stand, and how they are performing against your (reasonable) expectations. You can achieve this through a structured review system, but such systems often become ordinary formalities with little or no communication. The best time to give feedback is when the event occurs. Since it can impact greatly, the feedback should be honest, simple, and always constructive. If in doubt, follow the simple formula of:

- highlight something good
- point out what needs improving
- suggest how to improve

You must always look for something positive to say if only to offer some recognition of the effort, which has been put into the work. When talking about improvements, be specific: this is what is wrong, this is what I want/need, this is how you should work towards it. Never say anything as unhelpful or uninformative as "do better" or "shape up" - if you cannot be specific and say how, then keep quiet. While your team will soon realize that this is a formula, they will still enjoy the benefits of the information (and training). You must not stint in praising good work. If you do not acknowledge it, it may not be repeated simply because no one knew you approved.

Recognition and feedback are essential aspects to an employee's satisfaction in the workplace. It drives the creation of a positive work environment that reflects a respect for the individual; recognition of the contribution that individuals or teams have made to organizational success; the satisfaction from work and believing in the organization's mission; seeing results from our efforts, receiving feedback on performance, and having additional opportunities for learning. In short, what motivates is quality leadership.

⚠

> *"When you can't have what you want, it's time to start wanting what you have."*
>
> *- Kathleen A. Sutton*

Retention and Motivation

People issues are the number one cause of company retention issues. Even in the world of high technology systems running much of a typical organization's business, the motivation and performance of its employees can make or break organizational success. Ineffective understanding and communication with management or employees can be disastrous. The symptoms of some of these issues are very tangible and very expensive (i.e. High employee turnover, low staff performance, lawsuits, employee theft, low response to technical project efforts, and staff burnout).

These issues can be eliminated or dramatically reduced when they are understood and dealt with at the root cause level, rather than at the symptom level. Basically, people do what they are incentivized to do, not what the organization wants them to do. Let's say for example that your company wants to improve performance, so it says that overtime will be compensated at ten times the regular hourly rate. Surprisingly, people begin to work slower during their regular workday. Why? The incentive given makes it desirable for people to work late, but not necessarily to accomplish better results.

An organization that can create an energized, higher-calling environment will have higher retention and greater productivity. A larger purpose is the key to building loyalty, and companies with loyal employees reap the rewards. Many service organizations have long-tenured employees, because employees were able to easily link into their higher calling. A company can increase loyalty and decrease turnover through:

- Clear and frequent communication. When an organization lets employees know what is happening, employees feel more included and trusted.
- Continuous training and tuition reimbursement. The US Department of Labor's Bureau of Labor Statistics did a Survey of Employer-Provided Training of 1,000 companies in 2001. It showed that companies with high employee turnover train less than other companies. Providing training is important because it is an investment in employees, which they see as

money in the bank. When you invest in workers, they are more apt to invest in the company.

- Expect greatness. When you raise the bar, employees will meet your expectations and feel important.
- Provide career counseling. When you help employees grow in their careers, they are more apt to stand by you.
- Invest in employees' financial futures with a matching 401k. When you have a stake in their financial future, they will want to stand by you.
- Reward and recognize employees often. Employees crave positive feedback and will be more productive when they receive it.
- Ask employees for input on important decisions. Employees will feel important and more committed to the mission.
- Institute exit interviews with terminating employees to determine why they are leaving.
- Establish a family-friendly work environment. Child-care benefits and onsite ATM machines allow employees to spend more time with family. Allow employees to work on visible projects or add additional duties that interest them.

Loyalty to the department mission does not come easily. You must build it one employee at a time. And building loyalty is much like building trust: It is easy to tear down; the challenge is to build it up and maintain it.

A Manager's Role

Surveys have shown time and time again that the most important aspect of an employee's job is their manager. It ranks number one above work environment, benefits, and money. With such a strong supporting factor, one would think that managers would do everything possible to ensure that their employees are made to feel important. Unfortunately, most managers do not realize the strong influence they have over their employee environment.

As a manger, you must keep in mind that your emotions and actions set the pace each and every day. It sounds insignificant, but I guarantee you, if you are emitting the stress you are carrying, I am absolutely sure your staff is feeling it as well and creating a very tense environment. Projecting confidence is a critical aspect of a manager's position. IT managers are often so internally focused that they are unaware of the image they project. If they don't project confidence, they are unlikely to inspire and energize their group.

If there is one thing I have learned in life, it is the fact that everyone wants to be appreciated. This goes for managers as well as employees. Studies have shown that in today's business climate, rewards and recognition are crucial to staff retention. Although it may seem as though an employee is independent and self-sufficient, the fact is they need to feel valued. That is one trait that we never outgrow. Recognition for a job well done is the top motivator of employee performance. Yet most managers do not understand or use the potential power of recognition and rewards as a means of motivation or retention.

You must take the time to learn what makes each individual perform at their best. This is done through regular one on one's and staff meetings. Whether praising the work of an employee who has completed a project ahead of schedule, making a key hire, or dealing with a problem employee, you need to deal with staff issues as soon as they arise. In a business sense, technology seems like a bunch of back-office machines running the organization, but in reality, the people behind them make or break the operation. Therefore, it is even more important to ensure the technicians who service the customer are satisfied with their work environment. Doing your best to keep your employees happy and productive is a good investment of your time.

The bottom line is that good people will migrate to good work environments. If yours is not a good environment, you will lose your good people. So, here is my advice to IT mangers:

- Create the best possible environment within your department by reinforcing a culture of respect. To the extent possible, involve your staff in decisions that affect them. Give them as much control as possible over the projects and work they take on. Listen to their perspectives, and seek their advice and look for opportunities to recognize that their work is valuable to the company's larger agenda and mission. Seek ways for your staff to go beyond their job descriptions; advocate for them to participate in the new projects or task forces that will create your organization's growth.

- Continuous learning is key to employability. Establish relationships with individual members of your staff, and pay attention to what uniquely motivates different people. Ask them to identify what learning and skill-building experiences they are seeking to boost their viability in their industries. Make it possible for them to acquire this education or training.

The most sought-after professionals are constantly learning and updating their skills. If your company no longer pays for educational opportunities, try to negotiate time off for your staff to attend these courses.

- Give your staff opportunities to be identified outside the organization, and encourage them to become known throughout their industries. As much as possible, allow them to attend conferences and take on leadership positions within relevant professional associations, which will help them establish a national professional network, which is essential to employability. Though seemingly a counterintuitive approach to employee retention, your support demonstrates you are advocating for your employees' future, and this contributes to their satisfaction in their current positions.

Motivation

Your company's third round of layoffs is imminent, your boss is barking out demands with heightened urgency, and your staff is tired, tense and afraid they're next in line to be laid off. It's your job to motivate them, continue to meet ever-increasing demands and bring your projects in under budget and on time.

This all too familiar scenario is now a daily reality for many IT managers. Pressures from the top increase while resources and staff available to meet growing expectations decrease. The keys to effective management in this uncertain work environment lie in understanding the fundamentals that motivate people to perform, which means abandoning old incentives like job security and company advancement, and instead focusing on building a culture of respect.

At the core, individuals need to feel their work is valued and contributes in important ways to the organization's larger mission. This is no easy task, but nonetheless, a task demanded of all technology managers.

Recently, when making a reservation to fly on Southwest Airlines, I told the agent she had made my experience very positive. She seemed genuinely happy in her job, which is far from a highly paid or revered position. Her response: "We're just nice to people, and we try to help them," she said. "It's really basic."

Embedded in her simple response was a clear sense of her role within the larger team, and an understanding of the core principals that drive the company's approach to customer service. Not prescribed or based on fear that our conversation was being monitored for quality assurance, she understood her role was turning one piece of

Southwest's mission into a reality, one customer at a time. This scenario is not an isolated incident. I have a dear friend who works as a Sr. manager at Southwest's main hub. She personally confirms that Southwest employees are dedicated, trustworthy and happy due to a well-managed team environment.

A culture of respect and trust requires managers to build relationships with their staff based on honesty and integrity. It may be tempting to be outcome-oriented because of pressures to meet deadlines, but investing in staff relationships will yield greater results. Be sure to communicate both the positive and negative news you hold as a manager. This straightforwardness will build trust that you are an advocate for your area. To the extent that you can, involve employees in decisions affecting them and their work.

Another requisite for managers is to regularly communicate and prioritize their departments' goals, while tying accomplishment of these goals to the organization's larger aims. Even if company-wide recognition processes have been eliminated, effective managers find ways to recognize their workers' individual and collective accomplishments and convey publicly how their work has advanced the larger agenda of the department and organization.

Solid management in today's world of work requires managers to return to the fundamentals that motivate people, especially within the sphere they influence. Old rewards and incentives like trips to Hawaii have fallen to the wayside. Elaborate company celebrations have been eliminated. Key projects that ensure visibility and allow your staff to move up in the organization are no longer certain. Now you've been told there will be no salary increases this year. Your staff, many of whom have survived several rounds of layoffs, know that you cannot promise them what does not exist anymore - job security. These issues call for a return to a basic understanding of what truly motivates people: Meaningful work that has inherent value and is valued by others.

Money as an Incentive

While salary is a critical measure of one's success in an organization, and an undisputed necessity, its relationship to career satisfaction is overestimated. Money is important, however, it is less of a motivator than commonly thought. Individual merit incentives are shown to undermine teamwork, encourage employees to focus on the short-term, and lead people to link compensation to political skills and ingratiating personalities rather than performance.

The factors that retain employees go beyond salary. They include equipping your staff with the tools to be viable in an environment that demands self-reliance to navigate career advancement. There are some basic reasons why many top performers stay in positions even when they might earn a greater salary elsewhere, usually related to incentives you can provide your staff to create a more productive work environment while fortifying their careers.

IT Management Lessons

- Remember, from an IT department perspective you are only as good as your employees. When hiring, take your time and research each candidate's qualifications. The extra time invested will help you avoid finding out the hard way that you have the wrong person for your environment.
- Learn all you possibly can about the culture of your employees, your co-workers and the business. Cross-cultural awareness is key to managing a successful team.
- Focus on employee development. Maintain accurate, progressive yearly or bi-yearly performance reviews.
- Unfortunately, we cannot avoid employee behavioral issues. Maintain a written log of all conversations. Work to ensure that the behavioral issue does not affect the team.
- Let your employees know how you feel often. Maintain regular feedback or one-on-one sessions. Be careful not to confuse regular one-on-one meetings with overly needy employees.
- Reward hard working dedicated employees.

Part III - Creating Team Synergy

> *"A **good manager** doesn't try to eliminate conflict; he tries to keep it from wasting the energies of his people. If you're the boss and your people fight you openly when they think that you are wrong—that's healthy."*
>
> —*Robert Townsend*

Building Teams

In my opinion, teams are the hottest thing in IT! They have become an essential device for structuring job activities. But how do you as a manager create effective teams? The key components making up effective teams can be subsumed into four general categories: *work design, team composition, contextual influences and process variables.*

- *Work design* is extremely important to creating functional roles for team members. Teams work best when employees have freedom and autonomy, the opportunity to utilize different skills and talents, the ability to complete a whole and identifiable task or product, and a task or project that has a substantial impact on others. These characteristics enhance member motivation and team effectiveness because they increase members' sense of responsibility and ownership over the work, and because they make the work more interesting to perform.

- *Composition* includes variables that relate to how teams should be staffed: the ability and personality of team members, size of the team, member flexibility and members' preference for teamwork. To perform effectively, a team needs people with technical expertise, people with problem-solving and decision-making skills and people with good interpersonal skills. The most effective teams are neither very small (fewer than four or five) nor very large (approximately 12).

- From a *contextual influence* perspective, the three most important factors are adequate resources, effective leadership and a performance evaluation and reward system that reflects team contributions. Each plays an important role in facilitating effective participation by team members.

- *Process variables* include member commitment to a common purpose, establishment of specific team goals and a managed level of conflict. Members of successful teams put a tremendous amount of time and effort into discussing, shaping and agreeing upon a purpose that belongs to them both collectively and individually. Furthermore, when hiring for leadership positions, do not place too much emphasis on experience. Successful teams translate their common purpose into specific, measurable and realistic performance goals. These goals help teams maintain their focus on getting results. Conflict on a team is not necessarily bad. Teams that are devoid of conflict are likely to become apathetic and stagnant. Conflict can improve team effectiveness when it stimulates discussion, promotes critical assessment of problems and options and leads to better team decisions.

> *"Build for your team a feeling of oneness, of dependence on one another and of strength to be derived by unity."*
>
> *—Vince Lombardi*

Maintaining Productive Teams

Stimulating the development of high-performance teams is a top priority for many IT managers. Unfortunately, they often do not see themselves as active players in the process. Sometimes, this comes from a mistaken idea that a team should be self-contained and owned by the team members. It is a fact that a team is owned by its team members, however, it is the manager that plays THE KEY ROLE in setting the climate for the synergy and development of the team.

I cannot overstate this point. If you want to encourage team functioning, it is very likely that you, yourself will have to change. If you do not, any team approach is doomed to failure. If you look at teams in other contexts, you will quickly realize that leadership determines success. A sports team has a coach and a symphony orchestra has a conductor. These teams do not spontaneously develop without effective leadership, but develop and grow with the help and guidance of a leader whose job is not to control, but to teach, encourage, and organize when necessary.

A good way to describe the role of the manager is a catalyst, a force that causes things to happen for other people, and the team. Not only is the manager's role critical, but it changes over the life span of the team-building process. In the beginning of the process of team-building, the team members may need a good deal of help developing their mission and purpose, identifying what they want to accomplish, and, more importantly, help with the development of interpersonal and group skills such as conflict resolution, meeting management, etc.

You may also need to remind them that you are serious about the team, meaning that its activities and decisions or recommendations will be implemented wherever humanly possible. You may even be called upon to act as a mediator, when the team members cannot resolve conflict. As a team grows and matures, you *might* even become an equal team member.

Critical Leadership Factors

Listed below are some important leadership factors that are required in order to create a successful team building process. While these are particularly applicable to the formal unit leader (i.e. the IT manager), they also apply to team members who are performing in a leadership capacity. IT managers, or in this case team leaders, must:

- Recognize the importance of balancing between tasks (getting the job done) and people (ensuring that team members are satisfied with the process of getting the work done).
- Have highly developed inter-personal skills and understanding of some basic psychology regarding what makes people commit to, and perform.
- Have the willingness to listen and ability to communicate. Leaders must have a preference to listening and understanding rather than controlling and talking.
- Show constancy of purpose. Leaders must commit themselves to the team, and not give up when the going gets rough, or success is slow to come.
- Show consistency in behavior. Leaders must behave in a consistent manner regarding teamwork. Leaders who sometimes encourage team process and sometimes bypass the team confuse the hell out of everyone. When this happens, nobody takes the team or team goals seriously.
- Model desirable team behavior. The team will take its cues from its leader, or the manager. You cannot break inter-personal rules, not listen, and use autocratic prerogatives, and expect members of your team to believe that you REALLY value working together.
- Be able to deal with problem team members. Sometimes a team does not have the internal resources to deal with a member that is uncooperative or so unskilled in group behavior that he or she becomes a barrier. A manager must be able to coach when necessary, problem-solve, establish consensus and mediate.

The importance of teamwork is found in the functioning of a football team as one single unit. Simple stated, trusting one another is the epitome of teamwork, and it is just as true in the workplace as on the field.

Whether football or corporate politics, the fundamentals are:

- The team comes before any single player.
- Trust your team members because you must count on them.

- Team members must communicate clearly and consistently to other team members.
- Practice is critical.
- Never lose sight of each game's goal.
- Choose the best and most committed players for the team.
- Every player must give his all to do his very best in every game.
- After games, review what went right and wrong. Then make adjustments and do it better the next time.

Team Thinking

The leader must create the environment, and the team will build on its own. To create the conditions for your team to thrive, you must:

- Train your team members to understand how to be good team members with good communication skills.
- Provide your team with the resources it needs to generate ideas, solve problems and develop information.
- Help the team understand its purpose.
- Supply a good team leader to assist in guiding the team toward its purpose. Choose someone who can model trusting behaviors.

Reaching Success Together

The word that has the greatest influence on a team's success is "purpose". Ask your team to answer these questions:

- What are we trying to accomplish?
- How will we know if we have succeeded?
- What's the impact of failure?
- Who is the team's customer?

Teams must set ground rules to reduce conflict, increase productivity and improve effectiveness. A team can make as many or as few ground rules as it thinks necessary. Typically teams end up with five to 10 rules. A team can solve problems and come up with new ideas faster than any one employee can. Therefore, you should develop or mentor your team as a group during the team's formation, and reviewed and updated periodically. The key is in creating and nurturing your team to be the best.

It is a well know fact that the need to build effective teams is increasing and the available time to do is decreasing. So how do you increase team effectiveness in a climate of rapid change with limited resources? Here is an excellent team-building exercise that focuses on the feedback and follow-up process, which can significantly increase leadership and customer service effectiveness. This parallel approach has been shown to help leaders build teamwork without wasting time. It requires that team members courageously ask for feedback, have the discipline to develop a behavioral change strategy, to follow-up and to "stick with it."

To implement this process, you will have to coach or facilitate rather than be the boss of the project. Team members should develop their own behavioral changes, rather than have them imposed upon them.

1. Begin by asking each member of the team to confidentially answer two questions:

> A. On a scale of 1 to 10, how well are we working together as a team?
> B. On a scale of 1 to 10, how well do we need to be working together as a team?

> Calculate and discuss the results. On average, multinational corporations showed a team average of "5.8" level of effectiveness. The goal should be at least "8.7" overall.

2. Ask the team, "If every team member could change two key behaviors, which would help close the gap between where they are and where they want to be, which two behaviors should they all try to change?"

> Prioritize the behaviors and determine the two most important behaviors to change for all team members.

3. Have your team members choose two behaviors for personal change that will help close the gap. Then have them ask for brief progress reports from each other monthly. Progress can and should be charted.

> If team members regularly follow up with their colleagues, they will invariably be seen as increasing their effectiveness in their selected individual "areas for improvement." The process works because it encourages team members to primarily focus on changing their own behaviors.

⚠

"I have found that being honest is the best technique I can use. Right up front, tell people what you're trying to accomplish and what you're willing to sacrifice to accomplish it."

—Lee Iacocca

Remote Team Synergy

The challenges of maintaining a distance team include all of the challenges of building a team that works together in one site, plus the added variables of distance, time and culture. It's virtually impossible to micro-manage numerous groups and individuals who are located in different places, time-zones and cultures. To successfully build an empowered team, leaders must demonstrate the following abilities:

- *Create a shared sense of purpose.* When everyone shares the big picture about why the team exists, what it must do, when it will be done, and how results will fit in the structure of the various organizations, people can fit their contributions to other parts of the project.
- *Develop shared decision-making.* Everyone on the team needs to know their thoughts and actions are important to team success. All participants need to know the leader and other team members value their contributions. When decision-making is shared, high levels of commitment result. When high commitment permeates a team, people routinely pitch in to help each other if one falls behind or gets into trouble.
- *Build expected norms for behavior.* Team members must understand explicitly what they can expect from each other in terms of communication, support and respect. They must know what they will and won't tolerate from each other. When norms are clear, people can work together rapidly and correct mistakes rapidly because they trust each other.

Resolving Communication Obstacles

Difficulties in communication are the chief obstacles caused by time, distance and cultural diversity. When your team is spread across the globe it is critical that you employ the right technology to assist you in communication.

- Find out what forms of communication team members have available to them and how skilled they are with them.

- Learn everything you can about your teammates' work cultures. Find out how their projects typically develop: how they plan, how they communicate, what words like "plan," "communicate," "team," "review," "report," "project," "budget," and "complete" mean to them. What are their normal production horizons? What are their primary languages? Can you find interpreters? Can you find indigenous workers to assist you in understanding how differences in your teammates' cultures compare to your own understanding?

- Develop visual displays that convey meaning across cultures. Such graphic images can include time lines and flow-charts, but are not limited to these. They also include team identity and other symbols that might stand for a vision, such as clouds and sky; strategy, such as roadways or arrows; agreement, hand-shaking, and so forth. Wherever possible, design communication around such shared visual displays.

- Develop an information and communication plan that ensures every member always receives all information and distributions promptly. Ensure all sites have appropriate support and alternatives appropriate for their communication technology. Ensure that staff who distribute meeting agendas and materials have multiple methods of getting the information to recipients.

When you're the remote member attending a meeting by video-conference and everyone but you has had the materials in advance, no matter what the excuse, you can easily feel left out and less important. Don't let this happen to anyone on the team!

Team Communication Methods

Finding the right method for team communication should be carefully examined. The rule of thumb is to carefully match it to everyone's technology and communication needs. Furthermore, each communication method should fit the situation. For example:

- *Fly the team together to build or fix trust.* Flying the team should happen fairly early in the project or soon after orienting to the purpose and schedules. Use a travel budget to get everyone to the same location for a meeting of a few days in duration. Build social time and content into the agenda. Plan some team building activities along with project management activities. If people come from different cultures, provide activities through which the group can learn about each culture and each group. Plan a social function with a theme from each culture.

- *Fax to distribute graphics rapidly and for rapid coordination.* Usually, when all members can't handle common formats through email attachments.
- *Email to distribute important information and news.* To distribute documents as attachments, when members use compatible software and can handle common formats.
- *Video-conference* when you need to be reminded of each other's faces and expressions or to collaborate on shared images that people at one site may be sketching or correcting.
- *Share screens online* when people in different time-zones can extend the work around the clock because of time differences while updating and editing documents and databases.

The common thread or theme connecting these competencies for IT global managers is *connectedness.* People don't just need information and supervision. They need to feel and be connected to others in purpose and meaning. In developing your global leadership skills, remember business is personal. It's people, not just skill sets, with whom you are doing business.

Virtual Teams

An obvious case for paying attention to team emotional intelligence is with virtual teams. Communication can be challenging due to non-verbal and visual cues. This often results in interpersonal relationships that are more problematic. This lack of personal interaction within the team makes it extremely difficult to delegate and manage tasks, projects and issues.

Intensify the problems inherent in any team is the fact that virtual team members are often from different parts of the company, different cultures and even different countries. The challenges of working with diverse team members in virtual environments place even more importance on communication skills and emotional intelligence competencies. Therefore, there is a great need for building cohesiveness and commitment to a shared purpose.

The biggest issue with virtual teams involves communications problems. These issues fall into several categories: The first is lack of project visibility. Team members know what they are doing on an individual basis, but they are not always sure where their pieces fit into the whole puzzle. Second, there are sometimes problems in actually getting a hold of people, which makes it frustrating for team members to get a response from people as soon as they would like.

Occasionally, there are also constraints from the technology. Communication in a virtual environment has its own set of challenges. It's sometimes difficult to derive the meaning from text-based messages, especially if the person is attempting to be sarcastic or facetious. Guidelines on how to let others know the intention of your message, whether it's through the use of emoticons, is important. For the uninitiated, "emoticons" are those expressive little faces made out of parentheses, pound keys, percent signs, and so forth.

Here are some tips on alleviating virtual team communication problems:

- *Include face-to-face time when at all possible.* Have an initial meeting for the team members to get together, meet each other, and socialize. Meet face-to-face periodically throughout the life of the project or task involvement. These meetings will help to establish ties and relationships among team members. It is especially important in creating an effective working environment where the team members are interdependent.

- *Give team members a sense of how the overall project or task is going.* Send team members copies of the updated project or issue schedule and provide an electronic view of the project schedule on line using the Internet/Intranet. Charts can be published on the Internet using the team's Web site. The primary idea here is to improve the quality and type of communications with all team members. They need to know where they fit in the big picture.

- *Establish a code of conduct to avoid delays.* The code could include a principle of acknowledging a request for information within 24 or 48 hours. A complete response to a request might require more time, but at least the person requesting the information would know that the request will be addressed. No one likes to feel that his or her request has dropped off the radar.

- *Don't let team members vanish.* Use the Internet/Intranet or workgroup calendaring software to store team members' calendars. While this could be difficult to maintain on a daily basis, it should not be difficult to keep up with scheduled out-of-town absences such as vacations or business travel. Another approach is to agree that team members will let everyone know when they'll be going out of town. Electronic mail with a distribution list is both an effective and efficient way to avoid MIA's.

- *Augment text-only communication.* The Internet is a good place to store charts, pictures, or diagrams so everyone can have a look. The fax machine, although now considered old-fashioned, can help resolve many global communication issues.

- *Most importantly, develop trust.* Technology on its own is not enough to bring global teams together. Virtuality requires trust to make it work.

The issue of trust is at the center of successful virtual team management. The fact is that old-style command and control management is simply impossible in a virtual environment. As a leader you must learn how to change the nature of power and how it is applied to your environment. Virtual leadership is about keeping everyone focused as technology changes and old hierarchy structures deteriorate. Hence, your successful leadership on virtual teams will likely be determined how much of an expert you are on the matter at hand, not by corporate hierarchy.

Every manager faces specific performance challenges related to teams. However, virtual teams have a unique potential for delivering and failing on a larger scale. Therefore, you have to know when to deploy teams strategically, when they are the best tool for the job, and how to foster the basic discipline of virtual teams that will make them effective. By doing so, you create the kind of environment that enables the team, as well as the individual, to performance at their best.

Λ

> *"What sunshine is to flowers, smiles are to humanity. These are but trifles, to be sure; but, scattered along life's pathway, the good they do is inconceivable."*
>
> —*Joseph Addison, British Essayist, Poet, Statesman*

Team Morale

Even in the best of times, morale is a delicate, unpredictable thing. Will one employee sulk when another receives a promotion? Will a canceled project throw a team into a tailspin of accusation and apathy? Will a switch from one operating system to another cost you your best technician? Threading your way through these problems can be like negotiating a minefield. At any moment, something can blow up in your face and send productivity tumbling even as your employees commence to mumbling and grumbling.

Bad morale is insidious. Bad morale lurks and simmers just beneath the conversations in the break room. But if you keep your eyes and ears open, you'll know when it is there. (And) contrary to popular belief - It is not the Economy; it is management. Morale is more than a people issue; it is a business issue. Low morale increases turnover, and turnover (when unplanned) is bad for IT managers and their reputation, department, and efficiency. Low morale also causes declines in productivity and quality. To my knowledge, no figures exist to quantify IT performance declines, but the correlation between morale and business functioning is self-evident.

Stress and illness caused by excessive demands in work and personal life can seriously reduce a worker's productivity and have a direct impact on the bottom line. Those demands are magnified in IT departments, where maintaining morale can be singularly challenging. For example, as IT workloads continue due to layoffs, staff members may feel overloaded and confused. Adding to the dilemma is the fact that IT people tend to be introspective and technical in nature. They feel unappreciated because no one sees the creative effort and energy that goes into writing an application or completing a large technical project. And when IT staffers do receive feedback, it is usually negative.

Before the problem of morale can be tackled, a couple of ground rules need to be understood. There are no easy fixes or blanket solutions. Morale is not like a software program—there are no service packs or patches. It cannot be fixed in one day or one week, and it cannot be solved with free lunches, mugs, or other gifts.

What you need to understand about morale is this: The mood of your employees can be brought down by external factors, such as the state of the economy, but it is your leadership skills, or lack thereof, that will tip the morale scales one way or the other. In tough times, the people you are responsible for are looking for support, leadership and reassurance. If you ignore or underestimate that need, you will have a morale problem on your hands. Furthermore, lack of communication and bad management, or lack of confidence in management, are the two biggest causes of low morale. It does not matter what the economy is like, the issues may be closer than you may imagine.

The first step toward fixing bad morale is acknowledging that the problem exists. The second step is realizing that it is your responsibility to make it better. While morale may seem like the domain of Human Resources (HR), that is a cop-out. You are the head of a community, a family of professionals, and no competent IT manager would leave morale for HR to deal with. You have to be the one who sets the tone and defines the IT culture in any organization.

While IT manager's should not relegate responsibility for morale to HR, you can and should lean on HR for help. Having an HR representative participate in meetings can help make employees feel cared for by the company. The HR rep is also another person an employee can talk to. No matter how open your culture, employees are not always comfortable talking to management, and that means they may not tell the whole truth about how they feel if they do talk to you.

There are actions you can and must take in order to reverse a bad morale situation, including placing special emphasis on management basics such as communication, leadership and special programs for employees: training, rewards and recognition. Here are steps tailored for an IT staff that you can take toward recognizing and rehabilitating low morale.

- The warning signs of bad morale are often subtle and difficult to detect. Some old signals, such as increased turnover and frequent absences, aren't as reliable now as they once were. People are less likely to leave their position in the middle of a recession, when jobs are hard to find, and since Sept. 11 some companies have discovered that turnover has decreased as employees yearn for stability and security in their life.
- The biggest mistake you can make is to ignore the existence of a problem or rationalize it away. People should not be constantly uncommunicative. If they get negative in conversations, you have to perk up your ears. Look at their

faces. Are they laughing or smiling around the office? Are they quiet in meetings? If behavior and attitudes have changed, red lights should start flashing in your mind.

- Even if you have a large staff, you have to make an effort to get to know them, know their spouses, know what is going on in your staff's lives. I tell my staff all about my family because I care about them, and I feel they should know that I too have many of the same personal family challenges. I want them to care about me as well because I need a lot from them, and I need them to want to give that kind of effort.

- Having daily contact with staff is essential for maintaining morale. No organization is good at being proactive when it comes to morale. Most companies are reactive. The CIO will notice when productivity is down, but if you are just starting to notice, that means productivity has already been on the way down for six months or more. An IT manger's first line of defense when it comes to low morale is to listen closely. The clues lie in what employees are not saying.

- Look at the level of energy. Is your staff engaged and participating in projects and meetings, or are they withdrawn and lethargic? If people are no longer contributing, particularly if they used to speak up often, that's a sign. If people are talking but are being pessimistic, that's very telling. When things are good, IT workers feel like they can do anything. But when morale is down, they tend to feel like even one project is too much to get done. The sense of urgency on projects diminishes.

- Walk the corridors often, and stand at the door before a meeting starts and listen to what people say as they come in. You may think your employees feel comfortable discussing personal or professional issues that are affecting their state of mind, but you're probably wrong. You might have an open culture, but people learn from previous jobs to keep their mouths shut. Employees try to hide their feelings because they are afraid of a punitive backlash if they say anything. The best way to counteract that kind of wariness is to talk to your staff in a consistent, honest manner.

Honesty is the Best Policy

As stated before, the most important tool for recognizing and combating bad morale is communication. Even if your company is going down the tubes and the future looks grim, tell your staff what is really going on. The biggest mistake is to withdraw and keep information to yourself. There is nothing more miserable than when a manager holds up in his office and assesses a situation behind closed doors and does not share information with the staff. It is insulting to your staff's intelligence, and it is very destructive in terms of morale.

Being up front is important, as is honesty. You do not want to say everything is fine and then come back in a week and say the company is not going to make it. Manager's may sometimes keep information to themselves because they don't want to worry their employees, but even though they mean well, hiding information will only spark anger and distrust. Tell them what is going on, and do not hold back because you think it will worry or upset them. If you stay silent, you will not be protecting them. In fact, you will be leaving them vulnerable, and that will only make the situation worse.

As a leader, even if your own morale is down, you have to keep your head up. When morale is bad, it is more vital than ever that you be a fountain of energy and guidance, even if it is just by keeping a smile on your face. This is where your leadership skills come into play. Everything that you do or say or feel is projected to your staff, and even the slightest negativity will come across. (And) your staff needs that leadership.

However, building morale through improved communication means more than smiling and keeping an open door, it means developing trust between you and your employees, and that kind of connection is made by opening up about yourself, both personally and professionally. Tell your employees about your leadership style, your philosophy on work and life, the environment you want to create with them, your vision and how you plan to achieve it. Tell them how you want them to communicate with you, and tell them what your foibles are. Let them see you as you are. Let them know you have been through rough times before and how you plan to get through it this time. Thank them for staying with you through this. Solicit their concerns and offer your help in getting through obstacles. Get specific about how you intend to make the situation better if you can, and how you will support them.

One of the most powerful actions you can take is to involve the staff in addressing a morale problem. Hold a series of meetings with the goal of gathering suggestions for fixing morale. At each meeting, update your employees on what is being done and gather feedback on the process as it moves forward. By involving your staff, you give them some power over the situation, instead of them feeling powerless. It will make them feel like part of the solution, instead of part of the problem. By soliciting their advice, you demonstrate a level of respect and trust that they need to see.

Training Drives IT People

Another effective way of boosting IT morale and solidifying a sense of commitment between you and your staff is to boost up your professional development and training program. Training is especially important during down times when you are asking your staff to do more and, in some cases, take over unfamiliar jobs. Therefore, even in the toughest of times do not shut down your training program. If you do, you will find that a morale problem develops very quickly.

Even if your budget is tight, you can find economical ways to fund training. Options include online learning courses and self-teaching packages. Reimburse your staff for IT textbooks or create a library of books they can use. By maintaining the training program, you let your staff know that you are committed to helping them stay current. Training indicates an ongoing dedication to employees, and taking it away is one of the worst things you can do.

To sum it up, bad morale is a very real, very serious problem that demands good leadership. The first step is to acknowledge the existence of a morale problem. If you make the effort to examine your management and communication skills and address a need for improvement, morale will go up and you'll find yourself with a loyal, resilient staff that will not jump ship once the economy improves.

⚠

<u>IT Management Lessons</u>

- Steadfast communication is a vital key to building and maintaining your team.
- Watch for warning signs of team anomalies or moral issues. Catching the symptoms early will help avoid team function break down.
- Always include remote and virtual team members in every project or issue related meeting. This will ensure they always feel like part of the team, no matter where they are located.

- Be open, honest, and true to your team and they will loyally follow your lead.

Part IV: Technology

> *"It has been my observation that most people get ahead during the time that others waste."*
>
> *-Henry Ford, Founder of Ford Motor Company*

The Goal: 100 Percent Availability

It is common knowledge that technology is moving far faster than the ability of IT leadership to shift organizational vision, values, strategic goals, tactical goals, and behaviors. As a consequence, IT leaders must pay critical attention to transforming the organization while maintaining customer focus; ensuring that critical data center systems are continually available and user end applications are consistently stable. The implication is that availability is largely determined by how well designers, operators, and maintainers work together.

Consistent availability is the single most important objective for any company relying on systems technology for business. Every segment of the system's architecture, from operating systems and applications to hardware and communication links, must be designed with resiliency and fault tolerance in mind. Unreachable Web sites, crippled database servers, and extensive LAN/WAN anomalies can immobilize operations.

Availability is a function of whether a particular service is functioning properly. You can think of availability as a continuum, ranging from 100 percent (a completely fault-tolerant site that never goes offline) to 0 percent (a site that is never available). All sites, regardless of the business, have some degree of availability. For example, many of today's manufacturing companies target "3 9s" availability (99.9 percent) for their sites, which means that there can be only approximately 8 hours and 45 minutes of unplanned downtime a year.

Hardware failures, data corruptions, and physical site destruction all pose threats to a data center that must be available 100 percent of the time. You can enhance the availability of your site by identifying services that must be available, then identifying the points at which those services can fail. Increasing availability also means reducing the probability of failure. Decisions about how far to go to prevent failures are based on a combination of your company's tolerance for service outages, the available budget, and the expertise of your staff. System availability directly depends on the hardware and software you choose, and the effectiveness of your operating procedures.

To maintain a highly available site, you must understand potential causes of failure and take steps to eliminate them. The following list contains some of the more common types of failure and the elements that can cause and prevent:

- *Application software*: Create a robust architecture based on redundant, load-balanced servers. (Note, however, that load-balanced clusters are different from Windows application clusters. Commerce Server components are not designed to be aware of application clusters). Review code to avoid potential buffer overflows, infinite loops, code crashes, and openings for security attacks.

- *Climate control*: Maintain the temperature of your hardware within the manufacturer's specifications. Excessive heat can cause CPU meltdown and excessive cold can cause failure of moving parts, such as fans or disk drives. Maintain humidity control. Excessive humidity can cause electrical short circuits from water condensing on circuit boards. Excessive dryness can cause static electricity discharges that damage components when you handle them.

- *Data*: Conduct regular backups. In addition to regular backups, archive backups offsite. For example, you can archive every fourth regular backup offsite, to save space. If your data becomes corrupted, you can restore the data from backups to the last point before the corruption occurred. If you also back up transaction logs, you can then apply the transaction logs to the restored database to bring it up-to-date. Replay transaction logs against a known valid database to maintain data. This technique is also known as "Log Shipping to a warm backup server." This technique is useful for maintaining a disaster-recovery site (also known as a "hot site").

- *Electrical power*: Use Uninterruptible Power Supplies (UPS). However, because UPSs are typically battery powered, they are useful only for outages that last for short periods of time. Be sure to use a UPS that has the same power rating as your equipment. Use power generators as secondary backups to the UPSs. You can use generators for an indefinite period of time because they are fuel powered (diesel or gasoline) and you can refuel them if necessary.

- *Hardware*: Deploy redundant hardware components, such as redundant array of independent disks (RAID) disk arrays, disk mirroring, and dual disk controllers to minimize disk failures. Use a redundant disk controller. Use redundant fiber channel host bus adapters and switches (for SAN configuration). In the event of an adapter or switch failure, the backup adapter or switch provides an alternate path to the SAN environment.
- *Network*: Implement network redundancy with any combination of the following: multiple NICs, multiple routers, switches, LANs, or firewalls. Contract with multiple communications providers (i.e. MCI, Sprint, SBC, etc.) or set up identical equipment in geographically dispersed locations.
- *Security*: Contract an independent security audit firm to evaluate your environment. Deploy intrusion-detection tools. Deploy multiple firewalls.

Clearly, system availability percentages are dependant on the type of industry/operation. However, the 99.9 + per year range should be the norm for any organization you support. The following table defines typical availability measurements to help you decide what level of availability you need for your site.

Availability target	Seconds of downtime	Downtime per incident (assuming four incidents per year)
99.9999%	31.536 (approximately minute)	7.5 seconds
99.9990%	315.36 (approximately 5 minutes)	1.25 minutes
99.9900%	3153.6 (approximately 1 hour)	15 minutes
99.9000%	31536 (approximately 9 hours)	2.25 hours

Availability and Costs

Availability is a continuum that becomes increasingly expensive as you approach 100 percent availability. You must decide what trade-offs and compromises to make to fit your budget. Many firms mirror their data and applications at multiple locations. While multiple, redundant data centers can help reach the elusive goal of 100 percent availability, the cost of building and maintaining duplicate sites can be overwhelming. Constant monitoring of system performance is necessary to respond to any problems that arise. Thus, staffing and equipment to provide 24/7/365 coverage is also required.

The following information and table provides a sample framework to help you calculate the benefits of implementing failure prevention strategies. Note the numbers used are only a guideline. Use your own data and judgment to create a risk-assessment table for your own data center(s).

- *Likelihood of occurrence* is the number of times an error is expected to occur (from 1 to 10; the higher the number, the more likely the error is to occur). In the following table, the "O" column represents this value.

- *Detectability* is the ease with which a failure can be found (from 1 to 10; the higher the number, the harder the failure is to detect). In the following table, the "D" column represents this value.

- *Severity* is the degree to which the failure will affect the site (from 1 to 10; the higher the number, the more serious the failure and the more severe the outage). In the following table, the "S" column represents this value.

The table lists the examples of the types of failures that can occur and the effect of the failure, followed by a calculation of the relative probability number (RPN), using the following formula: *RPN = Likelihood of occurrence x Detectability x Severity*

Item	Function	Failure	O	D	S	RPN	Prevention	O	D	S	RPN
CPU (dual)	Application processing	Server might go offline	2	4	7	56	• Monitoring software • Remote-access software • Disable CPU • Restart	2	2	7	28
CPU (single)	Application processing	Server offline	2	4	10	40	• Monitoring software • Second CPU	2	2	8	32
Drives	Application and data storage	Server offline	5	4	10	200	• RAID 5 controller with hot spare	2	2	5	20

							• Monitoring software				
Firewall	Protection from intrusion and hacking	Information stolen, site altered, or site made inaccessible	4	4	8	128	• Monitoring software • Additional firewall	2	2	4	16
Load balancing	• Balance load on multiple servers • Enable automatic failover if a server goes offline	• All traffic goes to one server • Site inaccessible	4	4	8	128	• Monitoring software • Additional load-balancing	2	2	4	16
Memory	Application processing	Server offline	2	4	7	56	• Monitoring software • Additional server	2	2	7	28
NIC	Network connectivity	Server offline	4	4	8	128	• Dual NIC card • Monitoring software	1	2	4	8
Power supply	Power equipment	Site offline	4	4	10	160	• UPS • Monitoring software	2	2	2	8
RAID controller	Data storage	Server offline	2	4	10	80	• Monitoring software • Additional server	2	2	8	32
Server cluster	Data storage	Single-server failure, resulting in slower service	5	4	2	40	• Redundant servers designed into site	5	2	2	20

							architecture				
							• Monitoring software				
Switch	Connect to network	Some or all devices offline	4	4	10	160	• Redundant power supplies • Redundant connection cards • Redundant managemen t card • Monitoring software	1	2	4	8
Web server	Serve Web site application to customers	Site offline	5	4	10	200	• Additional Web servers • Load balancing	2	2	5	20

Keep in mind that no business operates in a static environment. Availability goals that are appropriate today may not be appropriate next year, next month, next week, or even tomorrow. System availability is as sensitive to the vagaries of the business climate as any other key performance measure. Business decisions that either enhance or impair availability are made every day at every level of the organization. Thus, availability goals must be reviewed and managed on a periodic basis to ensure sensitivity to business requirements and the efficient management of budgeted funds.

Risk assessment modeling and simulation should be used to engineer availability for any type of business. Use the availability model to continually monitor operational needs, improve maintenance and operating procedures, and to thoroughly understand the risk variables in relationship to achievable availability. Ultimately, this will help promote and improve bottom-line results throughout the life of the company.

Preventive Maintenance

Availability is a function of reliability and maintainability. In other words, how often equipment will fail and how long it takes to get the equipment back to full production capability. Reliability, maintainability, and availability are determined by the interaction of the design and preventive maintenance (PM) function, which is one of the most important aspects to ensuring availability. It is the only event where downtime is the exception to the rule. If everyone expects the airplane to fly efficiently and consistently, the plane must be brought into the hanger during its required maintenance window. (And) it is in the best interest of the business not to rush the maintenance progress. After all, no one wants to lose an engine in the air.

The difference between achievable and operational availability is the inclusion of maintenance support. Achievable availability assumes that resources are 100 percent available and no administrative delays occur in their application. Therefore, maximum operational availability theoretically goes to achievable availability.

Keep in mind that system anomalies or device failures are not always attributed to catastrophic events. Most recurring problems are often a combination of staffing issues, budget issues (such as needing additional hardware or software), and technology architecture issues. As IT manager, it is your duty and obligation to step in and help resolve these types of dilemmas. Your goal should be to focus on the issue and resolve the problem once and for all (and make sure it is fully documented!). Although solving specific technical problems should be the technical staff's responsibility, doing so keeps your team focused on new problems and your business running smoothly -- a true time-saver.

PM and system availability has become a mission critical operation and a significant source of revenue for many companies, including production operations, financial institutions, and e-commerce organizations. When any part of a site is unavailable, the company is most likely losing a great deal of money. As with any other technology PM concept, the key is to apply ways to reduce or eliminate downtime to the environment.

However, operating PM procedures also can have a significant impact on service availability. To avoid service outages, you must carefully consider service availability for all operating procedures. For example, communications companies in the United States typically target "5 9s" or 99.999 percent uptime (5 minutes and 15 seconds of unplanned downtime a year). All companies might attempt to strive for additional uptime for their sites, however, keep in mind that significant incremental hardware investments are required to get those extra "9s."

You can create an availability checklist to monitor the availability of your site. The availability checklist below provides some examples that your checklist should contain.

Item	Monitors
Bandwidth usage: per day, week, and month	• *Bandwidth.* How bandwidth is being used (peak and idle). You can use this information to project how much bandwidth you will need in the future. This will enable you to plan for the peak bandwidth you need for a peak business season, thereby avoiding problems associated with inadequate bandwidth. You can get bandwidth usage data from managed routers and Internet Information Services (IIS) 5.0 log analysis (by using the Commerce Server Data Warehouse). • *Usage.* How usage increases (whether it increases, when it increases, and how long it increases).
Network availability	Network Internet Control Message Protocol (ICMP) echo pings (available from most network monitoring software) compare your network availability to the level agreed to in your Service Level Agreement (SLA) with your Internet service provider (ISP) or data center environment. Improvement strategies should be applied if network availability falls below the level agreed to in the SLA. The formula for measuring network availability is as follows: (Number of successful ping returns/number of total pings issued) x 100%
System availability	• *Operating system.* Monitor normal and abnormal shutdowns of the operating system. The formula for measuring system availability is as follows: (Period of measurement-downtime)/period of measurement x 100% • *Microsoft SQL Server.* Monitor normal operation and failover events of SQL Server. • *IIS.* Monitor normal and abnormal shutdowns in IIS.
HTTP availability	• *HTTP requests (internal).* Monitor HTTP requests issued internally. Downtime occurs when the site fails to return a page or returns a page with an incorrect response. The formula for measuring HTTP availability is as follows: (Number of successful HTTP requests/ number of total HTTP requests issued) x

	100% • *HTTP requests (per ISP).* Monitor HTTP requests issued from ISP networks (such as AOL, Microsoft MSN, MCI, Sprint, and so on), to track whether or not users of the monitored ISP networks can access your site. • *HTTP requests (per geographic location).* Monitor HTTP requests issued from different geographic locations (New York, San Francisco, London, Paris, Munich, Tokyo, Singapore, and so on) to track whether or not users from respective areas of the world can access your site.
Performance metrics	• *Number of visits (per day, week, and month).* Monitor traffic information to assess the level of activity. Monitoring performance is not strictly part of monitoring availability. However, monitoring performance can sometimes provide advance warning about potential problems that can affect availability if you do not address them. • *Latency of requests for sets of operations and page groups (per day, week, and month).* Compare these metrics to your transaction cost analysis (TCA) test results to see how site performance compares to TCA predictions and to identify system bottlenecks. For more information about TCA testing. • *CPU utilization (per day, week, and month).* Monitor utilization on all servers and middleware. Group servers by function to more easily track and plan site capacity. • *Disk storage.* Group servers by function and monitor disk capacity (total disk capacity and free space). Review weekly and monthly history so that you can spot trends and plan for expansion. • *Disk I/O.* Group servers by function and monitor disk I/O throughput. Compare weekly and monthly history with the disk I/O rating provided by the manufacturer. If the observed I/O nears the disk I/O, consider adding more spindles (adding more drives to the drive stripe set) or redistribute disk I/O to multiple disk

controllers.

- *Fiber channel controller/switch bandwidth.* Monitor System Area Network (SAN) fiber channel controller bandwidth. (a SAN is typically used to interconnect nodes within a distributed computer system, such as a cluster. These systems are members of a common administrative domain and are usually in close physical proximity (a SAN is physically secure.) If the observed bandwidth nears the throughput rating provided by the manufacturer, consider adding more controllers and switches to redistribute traffic and get more aggregate bandwidth.
- *Memory.* Make sure that the amount of available memory is greater than 4 MB. If the system nears this level during peak usage, add more memory to the server.

There are multiple misconceptions about preventive maintenance. One such misconception is that PM is unduly costly. This logic dictates that it would cost more for regularly scheduled downtime and maintenance than it would normally cost to operate equipment until repair is absolutely necessary. This may be true for some components; however, one should compare not only the costs but the long-term benefits and savings associated with preventive maintenance. Without preventive maintenance, for example, costs for lost production time from unscheduled equipment breakdown will be incurred. Also, preventive maintenance will result in savings due to an increase of effective system service life.

Long-term benefits of preventive maintenance include:
- Improved system reliability.
- Decreased cost of replacement.
- Decreased system downtime.
- Better spares inventory management.

Long-term effects and cost comparisons usually favor preventive maintenance over performing maintenance actions only when the system fails.

There are many advantages for having a good preventive maintenance routine. The advantages apply to every kind and size of company. The law of PM routines is that the higher the value of assets and business equipment per square foot of company, the greater will be the return on a PM process. For instance, downtime in a financial call center operations environment can reach $10,000 per minute. A painful reminder of just how critical a part technology plays in any organization.

As mentioned earlier, preventive maintenance does involve some risk. The risk refers to the potential for creating systems issues of various types while performing the PM task. In other words, human errors committed during the PM task and infant mortality of newly installed components eventually lead to additional failures of the equipment on which the PM was performed. Frequently, these failures occur very soon after the PM is performed. Hence, the key to a successful Preventive Maintenance (PM) routine is scheduling and execution. Furthermore, priority should be given to all PM routines, along with a very aggressive process to monitor the schedule and ensure that the analysis is completed according to set schedules.

Λ

> *"Penny wise can be pound foolish"*
>
> *--Unknown*

E-Business Strategies

One of the most challenging aspects of supporting technology is how to help Sr. business management approach the investment of e-business technology for planning and meeting business model objectives. New computers, sophisticated Websites and database systems can be very expensive. In addition, users and management may be resistant to change and to learning new applications. However, the fact remains that the business will need and want to find ways to streamline their processes using electronic strategies. Furthermore, e-business strategies can and should be applied to your environment, which in turn also helps your technical staffs work more efficiently.

If you are considering implementing an e-business strategy (i.e. local Intranet, Internet B2B, or vendor communication.) in your area of support, the first thing to do is take an inventory of the capabilities that you currently have, what is working in your operations, and your limitations and frustrations. Then, without consideration of constraints such as cost or staff resources, list the things that you should ideally be able to do. Consider the following:

- What is the one thing that you are currently doing that is most valued by the business environment? (i.e. upper management, staff, the user community, etc.).
- What is the one thing that you currently do not do that the business environment wishes that you would?
- What is the one thing that would give you maximum competitive advantage? (i.e. budget, resources, etc.).

From a technology support perspective, look at the procedures and tasks that are currently absorbing staff time and resources. Is there potential to streamline these, or to recreate them in a way that would be more cost-effective? For example, many manufacturing organizations print large daily reports related to updated material requirements and schedule adherence. Numerous copies are printed out in order to ensure that all respective departments receive the updates on a timely basis (i.e. purchasing, planning, inventory control, distribution, and production). An alternative is to provide online, downloadable e-reports on your Intranet site. When the user enters their department link, they gain instant access to their materials reports in whatever format you choose (i.e. Adobe Acrobat (pdf), Word, html, etc.). The process can provide 24/7/365 data access service to the business. The process can help eliminate many of the previous maintenance and/or support issues for your team.

Unfortunately, the cost of such a project eventually must be addressed. (And) budgets have never been tighter than they are today. Therefore, it is necessary to invest sufficient personal time in conducting a proper assessment of the costs and benefits of an intranet for your specific organization. This is the best way of obtaining the ammunition you need to do the job properly, with an appropriate budget and a realistic timetable.

Cost and Benefits Analysis

It is not possible to conduct a meaningful analysis without a clear indication of what you want your intranet to achieve. There are many ways of categorizing what an intranet does.

- *Information publishing:* Using the intranet to deliver news and other information in the form of directories and web documents.
- *E-mail.* Implementing an e-mail system that integrates seamlessly with the intranet, allowing information to be both 'pushed' and 'pulled'.
- *Document management:* Using the intranet to allow users to view, print and work collaboratively on office documents (word-processed documents, spreadsheets, presentations, etc.).
- *Training:* Using the intranet to deliver training at the desktop.
- *Workflow:* Using the intranet to automate administrative processes.
- *Databases and other personalized systems:* Using the intranet as a front-end to organization-specific systems, such as corporate databases.
- *Discussion:* Using the intranet as a means for users to discuss and debate issues.
- *A connection anywhere, any time:* Using Mobile Intranet, your staff and the business can link onto the company's intranet wherever and whenever they wish. For example,

management can take part in a business contacts conference at the other end of the country.

- *Improved quality:* The manual data-updating process no longer being necessary, each staff member is certain he/she is working using up-to-date data, as everything is done in real time. Furthermore, the information is available anywhere, any time. For example: information on inventory quantities available or unavailable will be up-to-date. Therefore, there is no delay in delivery and customer satisfaction is ensured.
- *Better customer service:*   Since anything can be done online, there is no need to return to the office or contact a colleague to obtain technical information. Your technicians can have secure direct access (locally or remotely) to the intranet and can provide an immediate response to business requests. Your support customer service is noticeably improved.

Going forward, to make your cost and time investment calculations you will need to gather some basic facts and figures about your audience.

- The size of your intranet target population and, if appropriate, the proportion this represents of your organization's total projected intranet user population.
- The number of people within the target population who will need new PCs, the number who are currently not networked and the number who will be provided with access to the Internet.
- The average annual salary and benefits of the target population, the average working hours in a day and working days in a year (used to calculate labor savings and productivity gains).

Total up the benefits for each intranet category under the three benefit headings: direct cost savings, labor savings and productivity increases. Before making your calculations, it is necessary to determine the proportion of the target population that is affected by each of the intranet categories. For example, the whole population may be affected by the use of the intranet for information publishing, but only 30% for document management and 40% for workflow. If you don't make these distinctions, you are likely to over-estimate your benefits.

It is not very difficult to calculate costs. (And) if you are pretty creative you can reduce many of the expenses with in house development. However, it still takes a little more ingenuity to pin down the exact benefits. Based on my experience, there are three main categories of benefit:

- *Direct cost savings:* Savings in expenditure other than labor (print, paper, telephone, travel costs, etc.) that can be directly attributed to the introduction of the intranet. These can usually be calculated in three steps: (1) the number of incidences of expenditure in the time period, (2) the cost of each incidence and (3) the proportion of these that could be eliminated using the intranet. For example, if the number of pages of formal printed material received per person per year, the cost in pence per page was 500 (including printing and delivery) were 6p, the percentage of pages that could be delivered on-line would be 70%, the savings in pounds would be 500 x (6 / 100) x 70% x the size of the population.

- *Labor savings:* Savings in the amount of time required to carry out tasks as a result of introducing the intranet. These can be expressed in minutes per person per day. To calculate the saving, divide the number of minutes saved by the number of minutes in the day (60 x the number of working hours) and multiply by the size of the population and the average salary.

- *Productivity increases:* Increases in output per person attributable to the introduction of the intranet, expressed as a percentage. Because personal productivity has such a wide range of implications from job to job and B2B, it is easier to convert these to simple labor savings. For example, if the total productivity gains were 3%, calculate the savings as (3 / 103) x the size of the population x the average salary. The actual effect of higher productivity, such as increases in sales, will ultimately reflect a much larger estimate.

Depending on your operations, there are other technologies that might be appropriate for you to consider as you plan your future e-business strategy and investments. These could include handheld devices and laptops if you have employees who are often out of the office. Voice recognition software can be used for dictation by disabled staff, or those who are really not comfortable with keyboards. For larger organizations, virtual private networks may be appropriate for file sharing and synchronized database maintenance.

Merging your strategic needs together with your estimation of current inefficiencies in staff time and resources, you can create a detailed draft of the requirements. Then you can start to analyze the potential investments that you need to make, and the potential savings or revenue that you can realize in a short and long term basis.

⚠

> *"It has become appallingly obvious that our technology has exceeded our humanity."*
>
> -- *Albert Einstein (03/14/1879-1955)*

Security

One of the most important issues facing any IT organization today is security, an issue that has always been high-priority but because of recent events has now been catapulted to the top of the list of IT priorities. The demand for IT security professionals is at an all-time high, with salaries paid out to them increasing while those for others in the same industry going down for the first time in history. Therefore, in the case of information security, it is important that we recognize that:

- Information security is not a separate aspect of an IT operation, but rather is an integral part of it. It should not be treated as a special project that is distinct from all the other projects in an IT organization. All IT projects should have information security integrated as a quality or performance criteria, and likewise all activities for the sake of IT security should go through the same change-control procedures as other IT activities.
- Problems in information security will rarely get noticed before it is too late, and one cannot just wait for those major system outages or database violations before any action is taken. Information security more often than not will involve many small actions done on a consistent and regular basis.
- Information security is not just about preventing ill-intentioned human activities from the outside (a hacker violation), but includes security risks posed by intentional and unintentional activities.

The greatest threat to an IT infrastructure comes from within the organization it serves, from people who have been entrusted access rights into the systems they work with on a daily basis. If you use shared folders in your network and at one time or another have experienced lost files because someone inadvertently deleted it, you know exactly what I mean.

No matter how much investment is made in IT hardware, software, and facilities, high-availability can never be attained if the people who use these resources do not have the necessary disciplines to help maintain the good working order of the IT infrastructure. I will never forget this incident in a multinational healthcare company: the company financial database went down because someone had powered-off the server thinking it was not being used and he had wanted to make some efficiency changes to the server configuration. Needless to say, the situation negatively affected operations. Unfortunately, events like this ultimately reflect on the systems department as a whole.

In today's business environment, the exposure to people-problems is much higher because of the following realities:

- Many businesses outsource portions of their operations or business processes to outside organizations, which means increased reliance on people outside its direct management, culture, and control.
- The reliance of the IT infrastructure on public utilities means having to trust services provided by loosely regulated outside parties. For example, the use of the internet for communicating with suppliers and customers means using a public data exchange network that has many possible points of failures (i.e.the internet service provider, the telephone company, the power utility company, cable facilities provider, etc.).
- As your IT department is entrenched in the daily operational tasks of all employees from all departments within the corporate setup, the use of IT services is greatly increased and likewise the possibilities of human error introducing system problems.

The IT organization, during the mainframe years, ran a tightly controlled operation with most of the equipment residing inside a secured "glass house" facility where only qualified personnel were allowed entry. Access into the system, whether to run an application, extract data, or even generate a printout, was only allowed after an approval process that went through an average of 3 to 5 people. In those situations, secure the physical assets and you have most of the potential security problems eliminated.

That is not the case in today's cyber-connected world, where corporate data is scattered throughout multiple computers in widely separated locations and directly managed by most everyone in the organization. The physical boundary of old is no longer possible in a system, which allows access via wired and wireless means. Corporate data assets are all over the place, passing-through so many different hands that it is virtually impossible to keep tabs on where they are, who is managing them, and who is securing them.

Social Re-engineering

Much progress has been made in most all areas of the IT infrastructure, where computing power doubles every few months or so. I remember just twenty years ago when a 640kb personal computer was considered state of the art, and when having a second floppy diskette drive was considered a luxury item. But this exponential growth in the IT industry is clearly not visible with respect to user's skills, especially with respect to proper management of computing resources, which involves four main activities:

- *Education* - teaching the concepts, principles, and theory behind IT management, security, and high-availability.
- *Evaluation* - examining how current activities are performed versus the ideal situation or defined objectives.
- *Facilitation* - assisting people in the implementation of measures designed to help reach this ideal situation, whether through the introduction of new tools, processes, or organizational changes.
- *Feedback* - evaluating achieved results and determining adjustments needed to the implementation, achieving a closed-loop process when this feedback is used in the education phase above.

The response to many of today's security risks and data management issues is social re-engineering. Social re-engineering is a continuous process of improvement or rather, a continuous process of catching-up with the issues as they happen. It requires diligent education and re-education of people, keeping them updated of the changes to the system, threats to its stability, and required operational procedures. Please keep in mind that most of the time this work will be dominantly that of correcting false knowledge or information. For example, 7 years ago, a systems user came up to me and stated that I should be very careful not to get sick from the spreading of computer viruses. Bless her heart, she was a strong supporter of our department and was only looking out for our well being. Ridiculous as it may sound now, there continue to be many false beliefs regarding technology that circulate throughout the user community.

There are many areas where re-engineering people will greatly benefit the IT organization, but my focus here is in making sure that your people contribute to, and not undermine, the high availability of the IT infrastructure. The solution is consistent training of your staff to ensure that they are communicating the right information to the user community. In addition, make every effort to meet with your Sr. level business partners on a regular basis (i.e. weekly and/or monthly).

Techy Approach to Security

These days there are plenty of data security breaches to keep IT staffs busy. Regular reports of catastrophic viruses or security weaknesses challenge organizations that are becoming increasingly dependent upon computers, along with routine viruses, worms and hackers. Business critical information consists of customer files and transactions, strategic business plans and marketing strategies, budgets and other financial information.

A recommended safeguard is to use the "air gap" method: Physically isolating vital computers both from the Internet and from the internal network. This method can help protect key systems within highly secure departments. This will ensure that documents are not accessible via shared printers or drive shares. This may seem like an easy way out of securing a network environment, however, the confidentiality surrounding the data is much more critical than taking the chance on uncertain network security approaches.

If you work for a larger company you may prefer setting up a virtual private network (VPN) for highly sensitive areas such as Human Resources and Payroll. A VPN is a private data network that uses the public telecommunications infrastructure rather than private lines and employs security features such as encryption to ensure privacy. Another simple, but often undervalued, method of maintaining security is through tighter password control. The first step is to ensure use of "strong passwords," i.e., long passwords containing upper- and lower-case letters as well as numbers and punctuation symbols.

Hackers don't need to guess dates or pets' names to crack a password. They use software tools that rapidly run through all possible combinations until they find one that works. For example, a five-number password has 100,000 possible combinations, but a computer operating at more than two billion operations per second easily can crack it. On the other hand, a 10-character password using upper and lowercase letters as well as numbers and punctuation has 72 different options per character, or more than 37 trillion times as many as possible five-digit combinations. It is not impossible to crack a strong password, but it takes a lot longer.

Depending on the nature of your business, I recommend taking security a step further by purchasing software that does more than validate a user's ID. It also requires passwords of seven characters (five alpha, two numeric) that do not match the user logon. If the user makes a mistake when entering the password, the account is disabled for five minutes to prevent access by hackers trying random passwords. It also restricts user access. Employees can be restricted to logging on from their own workstations, and clients can only log on from certain Internet Protocol addresses as an additional protection against hacking.

Before venturing into a whole new security system strategy, I recommend you perform an internal audit process on your environment:

- Evaluate and determine the business risks and exposure.
- Ensure available vendor solutions are compatible with the company's existing software.
- Determine costs involved to buy, implement and upgrade the software.
- Identify training and staff commitments and assess existing controls including firewalls, routers, virus scanning, network logs and incident response plans.

Nowadays, if you talk about information security, you are considered obsolete. Most technology people fell you should talk of information assurance instead. What is the difference? Strictly speaking, there is none. Information security has been narrowly defined as that which pertains to protection from unauthorized access. In actuality, information security is all about ensuring that the right information is <u>always</u> accessible by authorized parties.

Information security as we now define it is the same as what they would like to call as information assurance. (And) you cannot address information assurance issues if the system on which this information resides is not available. Hence, the link between information assurance and high system availability. Therefore, you should look at the many people-related weaknesses that can negatively affect the security and high-availability of your IT infrastructure, whether due to human errors or ill intentions, from within and outside the organization.

Here is a summary of items to help you avoid IT security related system problems:

- *Install and maintain security patches.* Vendors such as Microsoft and others regularly issue patches to fix newly discovered security problems called "holes" in software already in use. Hackers distribute the code needed to attack the system by passing through the hole. The patch code closes the hole and protects the system from attack. An internal auditor should be responsible for ensuring that IT is aware of any new patches issued and installs them promptly.
- *Use simple technology.* If there is not enough staff to support a complex firewall, install a simple one. You can implement effective firewalls either in hardware or in software that provide great protection with little maintenance.
- *Monitor your network.* Network security has become a tremendous issue for most companies. Monitoring computer use logs, network logs, firewall logs, intrusion-detection-system logs and similar data sources require a lot of work to identify significant events. If the network operates on a 24/7/365 basis, it may be necessary to oversee the security and interpret alerts continuously.
- *Install appropriate user identification systems.* Authenticate the people who have access to systems either on-site or remotely. At the very least employees should select passwords to computer systems that are hard to guess and change them regularly. If a password is not enough protection, you should choose stronger techniques such as tokens (which range from devices plugged into the computer's USB port to credit-card-sized units that display passwords that change every minute) or biometrics. For companies requiring more sophisticated ID systems, fingerprint readers and iris scanners are becoming more practical and less expensive.
- *Account for invisible people.* Many companies require employees to sign a confidentiality/nondisclosure/computer-use agreement. But some with network access may not sign one, such as non-employees with only occasional access to their system, temporary workers, vendors or contracted

personnel. Another "invisible person" is the ex-employee whose access was not terminated promptly. Strong controls must assure IT knows when someone is leaving, so it can cut off that person's access immediately and make sure he or she doesn't have additional ways of getting into the system (for example, through another account or by using someone else's password).

- *Watch data backups.* A backup tape represents potential danger; it can contain confidential files in a form easily reloaded onto another computer. Employees should be particularly wary of what happens to the backup tapes when new ones are created and not throw them into a box of tapes to reuse. Until tapes are erased, they contain sensitive data and should be protected and inventoried. You should also ensure the credibility and security of your off-site tape storage provider.

- *Clear out long memories.* Files deleted long ago can still linger on the hard drive. When computers that processed confidential company or customer information are going out for repair, being returned to the leasing company at the end of the lease or being given to charity, it is not difficult to un-erase files. You should assure every computer going out the door, for whatever reason, is cleared of confidential information.

These basic methods of protecting data won't prevent damage from a major cyberterrorist attack, nor will they totally stop the most common security breaches. The most important thing to realize is that the field is constantly changing. Each new service provided or technology used offers its own challenges. And then there is the matter of staying one step ahead of the hackers. We will always have hackers who enjoy creating viruses and hacking into systems. Even though there is a great deal of technology to protect our vital systems, it's their hobby, it's their life.

IT carries the main burden of protecting an organization's data, but meeting these challenges requires close cooperation between business departments and IT. The biggest security threat is a lack of communication between business staff and the information services staff. As IT manager you need to bring the teams together and look at it on a monthly basis to keep up with changes in technology and business operations.

Security Risks of Wireless Networking

Wireless networking offers easy, cable-free connectivity for connecting to the Internet and other network resources. Increasingly, corporate users are installing wireless nodes on company networks to facilitate notebook computer connections, conveniently located network printers, and conference room connections.

In many cases, the wireless networking access points are setup without approval by the IT function, since wireless equipment is easy to set up and requires no special information to configure it for use on most networks. Although wireless networking products contain varying levels of security, most components have security disabled by default. When a wireless access point is installed in its out-of-the-box state, the corporate network is opened up to the public, bypassing any Internet connection firewall or other Internet security.

Unsecured wireless network access points allow access to your corporate network by someone who is not physically inside your office. Depending on the range of equipment, someone down the hall, on another floor, or even outside the building could access your network. Hackers can intercept data packets, gathering sensitive information by having access to your internal network. Sensitive corporate data is not the only (or necessarily the primary) interest to unauthorized wireless network users. High-bandwidth Internet connectivity is a high-demand commodity.

Hackers using a laptop with scanning software participate in "war driving", a method of locating open wireless networks in business and residential areas. Often the mapped results are posted on the Internet to let others know where they can find an open high-speed connection to the Internet. In addition to the bandwidth drain due to unauthorized use, there is a legal concern with the content that may be transmitted to the Internet using your corporate network. Transmission of illegal copies of digital media and other inappropriate material could put your company at risk.

Protecting your Network

There is virtually no way to completely secure wireless networks. The most you can do is minimize and control the access rights that a wireless network has in reference to the corporate network. You should treat wireless networks with the same degree of security as you treat SOHO networks for employees working remotely. There are various tools for helping to secure a Wireless network including intrusion detection systems, authentication systems, and encryption systems.

A good first step would be to publish a corporate policy on wireless networking. Let employees know that they will be required to activate the security features of the wireless networking equipment. Securing the wireless devices is the best first step to protection as encrypted transmission is a standard feature of most wireless components. In addition, scan for open wireless network access points on a regular schedule. Design standard network segments for remote access and use perimeter security policy to reduce any liabilities. Then setup auditing equipment to aid in tracking down any potential compromises.

Wireless networking offers great convenience, but with the convenience can come great risk. Take the time to secure your wireless networking components and become aware of what wireless network access is available on your network. A little effort will go a long way towards reducing potential security problems.

Λ

> *"Any sufficiently advanced technology is indistinguishable from magic."*
>
> *--Arthur C. Clarke*

Contingency Planning
Disaster Management

There are no certainties and guarantees in life (except perhaps for birth, death, and taxes!). The unexpected can strike, with devastating consequences for our activities, with little or no prior warning. Any disaster recovery plan will largely be dictated by the agreement you have made with your customers, the extent to which they depend on the service you provide, and the cash you have available to spend on a back up strategy.

Disaster recovery has risen towards the top of the agenda in recent months, because of the concerns around cyber-terrorism. There are three elements to a disaster recovery:

1. Reducing the likelihood of disaster (often called redundancy)
2. Contingency plans to restore partial or full service as soon as possible after a disaster has occurred
3. A communication strategy to keep all those involved (but most importantly, the customer) fully informed of the current state of affairs.

The most common catastrophes are likely to be:

- Fire
- Flood
- Terrorist activity
- Earthquake (in some parts of the world)
- Severe structural damage to buildings from a variety of causes
- IT systems failure
- Key staff absence (due to accident, illness, or industrial action)

The consequences of the above can be very expensive, in terms of the replacement costs, loss of business, and time lost, or wasted, for customers. The objective here is to protect the information service from harm through improved operating and safety procedures.

First and foremost, a risk assessment should be carried out. The risk assessment should identify potential threats to your service. These might include:

- Electrical problems
- Lack of adequate back-up for computer systems
- Leaking roofs or windows
- Dampness in rooms where valuable documents are stored
- Fire hazards

Once the assessment has been carried out, you will be able to separate those areas where you can take immediate action and those where you will have to play it long, most probably seeking help and cooperation from other areas of the organization.

In most cases, the most immediate action would apply to areas such buildings and security. For example:

- *Buildings* - Fire prevention and safety is of vital importance. If you have no organizational plan, develop one in consultation with in-house or external experts. Schedule regular safety inspections to cover buildings, electric, fire prevention, computer security and reliability, and any other aspects of your set up which you have identified as having the potential to render your service vulnerable.
- *Security* - A common cause of problems is vandalism. Again, if you are not covered by an organizational security strategy, you need to devise one for yourself. Security is defined as 'The protection of assets of all kinds against loss from fire, theft, fraud, criminal acts, or other injurious sources.' The first step in setting up a security strategy is to contact your local security department manager.

Once you have identified the risk areas, and formulated a strategy for avoiding disasters, the next step is to plan for business continuity, should a disaster occur. This should ideally be turned into a handbook for all involved, on disaster prevention and business continuity, for distribution to key staff. You may wish to have two handbooks, one for prevention, and one for continuity. It is important that current copies of this handbook are kept in a secure place off-site, ready for distribution should disaster strike.

The continuity handbook should include, or cover the following:

- Floor plans, including fire exits, utility turn-off points, location of fire extinguishers
- Emergency evacuation procedures
- How to deal with:
power failure
severe storms
flooding
fire
earthquake (in those areas vulnerable to same)
terrorist activity and civil disorder
explosion
- List of personnel to contact (list must be kept up-to-date)
- List of facilities providers (data storage; equipment vendors and transport services; freeze or vacuum drying facilities; disaster recovery or business continuation specialists)

The information on your computer system may well have a much greater value than the equipment itself, even if you are a small operation. That information can easily be destroyed, but there are measures which you can take to ensure that even if disaster strikes, you can continue to access that information and run your business.
The questions to ask yourself are:

- What would happen to your service, and its reputation, if you lost your computer system for an hour, a day, and a week?
- What systems and records are you legally required to maintain, what would be the consequences of irretrievable loss?
- What are you contractual obligations towards, and service level agreements with customers, and what would be the consequences of breaking these?

The most common threats to computer information are:

- accidental erasure
- hardware failure
- Software failure

The solution, to a certain extent, is simple, backup, backup, backup. A daily local tape backup is the most common form of data backup to implement. Daily backups provide a good compromise between price/practicality and the necessity to recover data. Although this is comfortably the position of many who work with critical data, it does not necessarily provide the data accessibility that most expect from their disaster recovery solution. The question is how quickly can you be up and running again with your critical data after a disaster? If a critical component in a server dies and it takes a day or two to get replacement parts, the speed of recovering from your daily tape backup becomes irrelevant.

In most cases an Internet-based data backup and recovery solution can help address this concern directly by offering not only off-site, managed backups but also rapid recovery of critical data. Data is backed up over a high-speed Internet connection to the company's servers. Depending on your needs, you can backup only your most critical data, usually user data, or backup your entire server with a more comprehensive service plan. The backups are monitored in real time for success. And if data is lost, the backup provider can push the data back down to your server over your Internet connection, minimizing restoration downtime. In the case of a dead server or location disaster, such as a fire or flood, your data can be restored to an alternate server even at an alternate location.

Off-site storage of any kind, Internet-based or traditional, greatly improves the disaster recovery capability; however, storing tapes off-site still can be inconvenient. Retrieving a tape, much less the data from a tape stored off-site can take an unacceptable amount of time in today's business environment.

For those of you who already have a disaster recovery plan...Test it! Not only to make sure that the backup process is functioning, but to test how you will access your data if your server becomes unavailable. The goal is to answer the important question: "How long will it take from the point of disaster to have full access to our data?"

Power Protection
The concept of power protection speaks for itself. Hence, it is essential that you protect every aspect of your technology environment. These areas include:

- *Peripheral Equipment:* Power spikes and surges can destroy your IT investment instantly. Modems, fax machines, laser printers, monitors and other peripheral pieces of your workstation are all susceptible and all require protection from the common power fluctuations that can damage electrical equipment.
- *Desktops:* Many vendors such as Dell and IBM build in tolerance for the continuous onslaught of power anomalies your PC is likely to face. In most cases, you may not even

be aware of the frequency with which power problems hit your computer because of the protection already built in. However, the IT equipment does need protection against out-of-bounds power events with quality surge protection and reliable battery backup power. The designer may have accounted for the normal range of problems, but high-level power spikes or low-level power slumps are beyond the responsibility of the designer. Here the responsibility for protection rests with the equipment owner. When a workstation is connected to the network the importance of data protection extends past the desktop. In network environments, any data corrupted locally on a workstation could be written back over the unaffected data on the file server, causing lost data and unknown expense to the company.

- *Servers:* The most dangerous seconds in a file server's life occur when it is writing to disk. A power problem, be it a spike or a brownout, at this time can be catastrophic. If a power problem occurs while your server is saving, you could lose an entire hard disk. Servers used to write to disk, are at increased risk of data loss or hard disk damage due to a power problem.

- *Telecom:* Each segment of your telecoms network is susceptible to power problems. These range from lightning strikes to complete blackouts or power failures. To ensure total protection of a telecoms network it should be protected from customer service PC's to PBX and Key System Applications. A power failure at an unprotected telecoms network location could mean the disconnection of hundreds of phone calls and hundreds of dissatisfied customers.

Business Continuity

One of the most critical responsibilities of the IT manager position is to maintain an organized disaster recovery plan and ensure the readiness of the technical staff in the event a natural, technical or man made disaster. The primary objective is to protect the organization in the event that all or part of the operations and computer services are rendered unusable.

A disaster recovery plan is a comprehensive statement of consistent actions to be taken before, during and after a disaster. The plan should be documented and tested to ensure the continuity of operations and availability of critical resources. The planning process should minimize the disruption of operations and ensure some level of organizational stability and an orderly recovery after a disaster. Other objectives of disaster recovery planning include:

- Providing a sense of security
- Minimizing risk of delays
- Guaranteeing the reliability of standby systems
- Providing a standard for testing the plan.
- Minimizing decision-making during a disaster

The methodology behind the plan requires the following:

- *Obtain Top Management Commitment.* Top management must support and be involved in the development of the disaster recovery planning process. However, you should be responsible for coordinating the disaster recovery plan and ensuring its effectiveness within the organization. You should commit adequate time and resources to the development of an effective plan. Resources could include both financial considerations and the effort of all personnel involved.
- *Establish a planning committee.* You should appoint a planning committee to oversee the development and implementation of the plan. The planning committee should include representatives from all functional areas of the organization. Key committee members should include many of your top level IT managers or leaders. The committee also should define the scope of the plan.
- *Perform a risk assessment.* The planning committee should prepare a risk analysis and business impact analysis that includes a range of possible disasters, including natural, technical and human threats. Each functional area of the organization should be analyzed to determine the potential consequence and impact associated with several disaster scenarios. The risk assessment process should also evaluate the safety of critical documents and vital records. Traditionally, fire has posed the greatest threat to an organization. Intentional human destruction, however, should also be considered. The plan should provide for the "worst case" situation: destruction of the main building. It is important to assess the impacts and consequences resulting from loss of information and services. The

planning committee should also analyze the costs related to minimizing the potential exposures.

- *Establish priorities for processing and operations.* The critical needs of each department within the organization should be carefully evaluated in such areas as: Functional operations, key personnel, information, processing Systems, service, documentation, vital records, and policies and procedures. Processing and operations should be analyzed to determine the maximum amount of time that the department and organization can operate without each critical system. A method of determining the critical needs of a department is to document all the functions performed by each department. Once the primary functions have been identified, the operations and processes should be ranked in order of priority: essential, important and non-essential.

- *Determine Recovery Strategies.* The most practical alternatives for processing in case of a disaster should be researched and evaluated. It is important to consider all aspects of the organization such as: facilities, hardware, software, communications, data files, customer services, user operations, MIS, end-user systems, and other processing operations. Alternatives, dependent upon the evaluation of the computer function, may include: hot sites, warm sites, cold sites, reciprocal agreements, two data centers, multiple computers, service centers, consortium arrangement, vendor supplied equipment, and combinations of the above.
Written agreements for the specific recovery alternatives selected should be prepared, including but not limited to the following: contract duration, termination conditions, testing, costs, special security procedures, hours of operation, specific hardware and other equipment required for processing, personnel requirements, guarantee of compatibility, availability, priorities, and other contractual issues.

- *Perform Data Collection.* Recommended data gathering materials and documentation includes: backup position listing, critical telephone numbers, communications Inventory, distribution register, documentation inventory, equipment inventory, forms inventory, insurance Policy inventory, main computer hardware inventory, master call list, master vendor list, microcomputer hardware and software inventory, notification checklist, off-site storage location inventory, software and data files backup/retention schedules, telephone inventory, other materials and documentation.

It is extremely helpful to develop pre-formatted forms to facilitate the data gathering process.

- *Organize and document a written plan.* An outline of the plan's contents should be prepared to guide the development of the detailed procedures. The outline can ultimately be used for the table of contents after final revision. The disaster recovery plan should be structured using a team approach. Specific responsibilities should be assigned to the appropriate team for each functional area of the company. There should be teams responsible for administrative functions, facilities, logistics, user support, computer backup, restoration and other important areas in the organization. The IT management team is especially important because they coordinate the recovery process. The team should assess the disaster, activate the recovery plan, and contact business management. The IT management team also oversees documents and monitors the recovery process.

- *Develop testing criteria and procedures.* It is essential that the plan be thoroughly tested and evaluated on a regular basis (at least annually). Procedures to test the plan should be documented. The tests will provide the organization with the assurance that all necessary steps are included in the plan. Other reasons for testing include: determining the feasibility and compatibility of backup facilities and procedures, identifying areas in the plan that need modification, providing training to the team members, and clearly demonstrating the ability of the organization to recover.

- *Approve the plan.* Once the disaster recovery plan has been written and tested, the plan should be approved by Executive management.

Disaster recovery planning involves more than off-site storage or backup processing. You should also develop written, comprehensive disaster recovery plans that address all the critical operations and functions of the business. The plan should include documented and tested procedures, which, if followed, will ensure the ongoing availability of critical resources and continuity of operations.

The probability of a disaster occurring in an organization is highly uncertain. A disaster plan provides a certain level of comfort in knowing that if a major catastrophe occurs, it will not result in financial disaster. Insurance alone is not adequate because it may not compensate for the incalculable loss of business or data during the interruption.

Other reasons to develop a comprehensive disaster recovery plan include but are not limited to:

- Minimizing potential economic loss
- Decreasing potential exposures
- Reducing the probability of occurrence
- Reducing disruptions to operations
- Minimizing insurance premiums
- Reducing reliance on certain key individuals
- Protecting the assets of the organization
- Ensuring the safety of personnel and customers
- Minimizing legal liability

IT Management Lessons

- Be fully aware of all costs associated with any type of system downtime.
- Maintain a failure prevention matrix. This will ultimately help prevent any costly downtime.
- E-business strategies are great for improving efficiencies and streamlining processes. However, make sure you have sufficient needs analysis data. This will help avoid costly and needless e-strategy projects.
- Always maintain a security conscious approach to every technology project and implementation.
- Disasters and hardware failure are inevitable. Maintain a well organized and detailed contingency plan for every aspect of your environment.

Part V: Communicating with the Business

> *"The fewer data needed, the better the information. And an overload of information, that is, anything much beyond what is truly needed, leads to information blackout. It does not enrich, but impoverishes."*
>
> *--Peter F. Drucker*

Communicating Systems Issues

The scenario within every organization is one of common knowledge, a critical server needs to be rebooted on occasion and no one in the business knows exactly why. Similarly, at least one key process requires a series of manual maneuverings and incantations before it can be revived after crashing. Often, the fixes for these kinds of anomalies remain inside the heads of the systems or engineering staff, and once resolved, staffers go about their business. Furthermore, the cause or fix remains unannounced due to weak communication links between IT and the business. From a business standpoint, this lack of communication can create an unstable support and availability image for your department. Therefore, communicating systems events that may affect the business (i.e. unplanned failures and anomalies) can be the most important yet challenging aspect of an IT manager's position.

There are all kinds of models of communication, some basic and some complex. For technology purposes communication can be described as creating understanding. Through words, actions, body language, voice tone, and other processes you send many messages about yourself, the issues, and your technology organization. This constitutes precisely one-half of the communication process. The second half consists of verifying that the message you intended to send was actually received and interpreted the way you intended. The only way that you can be sure you have created understanding is to listen to the people you are communicating with, and make special effort to encourage them to reflect back to you what they have heard (and what they make of it).

When business is interrupted by a technology issue, it usually means a loss in productivity ($$). (And) a loss in productivity means that management will need to have a clear report or understanding of how and why the situation occurred. Often you may have the exact answer immediately available and other times you may need more time to research the issue. Regardless of the circumstance, you will still need to provide a clear explanation of the cause, resolution or status. Therefore, always keep in mind the following when communicating to business management:

- Although you communicate in a way that seems clear to you, the receiver of the communication filters the information through a very complicated set of pre-conceptions that can function to distort the message received. Relaying the technical aspects of an issue is a perfect example. You must gauge your audience and refrain from being too technical during business management meetings.
- Receivers listen selectively. They hear and process some things and gate-out other things. That means that while you may have explained the "whole picture", is it likely that the whole thing was not received.
- The ONLY way you can ensure that you have created common understanding is by asking the other people what they have heard, and what their reactions are to it. Again, during technical meetings related to production interruption, ask and ensure that the receiver understands exactly how and why they were affected.

Planned Outages

It is important that the message(s) your systems department communicates to the organization be clear and unwavering. Obviously it is also important that your department be as flexible as possible. However, the consequences of pushing a process such as an All Systems Outage (ASO) or IPL can be far more detrimental than the time the business is attempting to save. Therefore, if systems cannot meet the business need, the business will have to adjust their schedules accordingly and without reproach.

The biggest challenge is finding the right communication method. As mentioned before, there is no real substitute for person-to-person communication. However, for situations such as IPLs, ASOs, or planned outages to replace hardware, the most effective method of communicating the details would be via email. This is especially applicable when large numbers of people need to know the details of the situation.

E-mail can benefit you in several ways:

- Communicate details to a large number of users simultaneously

- Allows management and users to respond, inquire or communicate concerns
- Provide continually updates regarding outage status
- Provide schedule changes or updates
- If necessary, deliver task documentation for user related processes (i.e. instructions or training material)
- Send system to ensure online vendors are aware of the process

Regardless, of what you have planned that day, always make time for your business customer's (i.e. management, users and vendors) questions and/or concerns. The success of your technology leadership and your staff largely rests on the effective contact you maintain with the business.

The following guidelines can help you create effective email responses:

- Any type of downtime (planned or unplanned) immediately drives concern within business. When receiving an email response from a concerned or upset customer try to focus on the business aspects of the situation. Analyze the email thoroughly before responding. Reciprocate for survival.
- Communicating by e-mail is no different from writing on your company letterhead. A business communication is business, period. A certain degree of formality is required. However, do not take negative connotations personally. Just because e-mail tends to be more immediate and personable, it does not need to get personal.
- If your correspondent uses your first name, then by all means use his. Many people do not want such immediate informality in a business situation, especially in the international arena.
- Respondents frequently try hard to be brief. This is desirable, but business messages are usually longer than personal notes. It is important to communicate and not kill understanding with brevity.
- When replying to long messages you will often be replying to only part of the received message. Save space by not returning the whole message, only the part to which you are replying.
- Short, plain sentences are easier for someone reading in a language other than their mother tongue. In some parts of the world, the written language is very formal and quite different

from the spoken language. Therefore, there is an expectation that your written communication will be formal. For international business e-mail, err on the side of caution and write in a formal tone. It is easy and natural to progress from formal to friendly, but it weakens your position to have to step backwards from friendly to formal.

- Get to the point and avoid lengthy emails if necessary. In problem situations do not attempt to confuse but rather make your message short and clear. You risk losing the effectiveness of your message if it is too long. Keep emails succinct.
- Plan what you want out of the situation before you go into it.

⚠

"The politics of business is more difficult than physics."

--Albert Einstein

Communicating Systems Projects and Changes

Even more important than day-to-day or week-to-week issue communication is the message you convey for large-scale projects or changes. It is important that when you communicate the details that you know what kind of messages you wish to send, and what you want people to take away from your communication.

Whenever you communicate to the business about change, you should be striving to convey the following position:

- That you are personally committed to the change and seeing it through, even if it has negative consequences.
- That you recognize that the change may impact production.
- That you are open to discussion of the feelings of management regarding the change.
- That you are confident that the "team" can make it through the changes.
- That you want and need input to make the changes go as smooth as possible.

As a systems manager you need to make decisions about who you must communicate with, what needs to be communicated, when you will communicate and how you will do it. Most technology managers sometimes have a tendency to communicate about change on a "need to know basis". To be an effective manager you must recognize that almost any major systems change will have effects on most users in an organization, no matter how removed they are from the change.

Except for situations that involve confidentiality, users who are indirectly affected will likely want to know what is going on, and how it may affect them. This applies to your own staff and those organizations that are related to you (i.e. other technology groups and client organizations). The basic rule of thumb is that communication should take place directly between yourself (the IT manager) and upper business management. The message should then be disseminated to lower level managers and employees by Sr. management. (And) depending on the impact to the business, you should schedule daily, weekly or monthly meetings to discuss progress status and concerns.

If you need to determine what to communicate, keep in mind what you are trying to accomplish through your communication about projects or change. When you communicate you are trying to:

- Give information that will reduce uncertainty and ambiguity regarding the change.
- Pre-empt the hidden information system of the grapevine, so you can ensure that incorrect anxiety provoking information is not spreading.
- Provide forums for employees to communicate their reactions and concerns to you.
- Communicate as much information as is available to you. Obviously, you need to exercise judgement where there is confidential and/or sensitive information involved, or where your information may be unreliable.

The longer you wait to communicate details of an issue, change, or project, the more likely you are to extend the period of adjustment. This is because it is very difficult to "keep a lid" on anything in technology, and even if you are silent, your staff will likely hear vague things through the grapevine. Grapevine information tends to be sketchy enough that it creates a high degree of anxiety, and also a high degree of mistrust of management.

Therefore, the earlier you communicate the less likely erroneous or upsetting information will come through the grapevine. Communicate as early as possible about projects, but do not assume that once you have done this the job is over. Communication should occur in anticipation of change, during the implementation, and after the project or issue has been stabilized.

Another decision you need to address is what needs to be communicated in group settings, and what needs to be addressed in one-on-one meetings with management or staff employees. However, most situations require both group communication and one-to-one communication. They compliment each other. Using only one or the other will create problems. Below are some guidelines.

Use group communications if:

- You need to ensure everybody hears the news at the same time.
- You want to encourage group discussion to generate ideas and the problem solving process.
- You want to increase the sense of team synergy.

- You wish to set the stage for individual meetings. For example, for project task assignments, you can call a short group meeting to announce the goals generally, then immediately meet individually with each staff member to inform them of the details.

Use individual meetings if:

- The changes are likely to cause a high degree of emotionalism that is better dealt with in private.
- You want to ensure that shyer people have a chance to express themselves.
- The changes involve elements that should remain confidential (payroll system changes, financial related implementations, etc).
- You need to have detailed discussion about the change with specific people.

Communicating with Difficult People

It is inevitable that you will on day experience the challenging effects of a difficult user or manager in your career. The effects of their behavior can range from minimal personal frustration to career affecting issues due to failed projects. For the difficult person, it is not whether someone is right or wrong, but rather it is about making things difficult for everyone.

Difficult people will yell, explode, and try to intimidate you by threatening to go above your head. It is a rare occurrence to hear from someone who is free from these hostile and manipulative people. However, the probability of encountering these people is extremely likely in every large organization. According to research, difficult people make up only 3-5% of the population, yet they create over 50% of the everyday problems.

Certainly, we all can be miserable, hostile and basically pretty unpleasant at times. But difficult people are this way all the time. A brief encounter with a difficult person leaves one angry, frustrated, and demoralized. These people go right for the jugular vein. The negative behavioral patterns they learned are used strategically to wear you down. Their only objective is to challenge your objectives and win regardless of who stands in their way.

There is no full proof way of eliminating the negative affects of difficult people. The best thing to do is cope with them and work with the situation to the best of our abilities. The first step in coping with a difficult person is to understand why they behave this way. Generally, these people are unhappy, insecure, and have low self-esteem. In some cases it may be personal due to a work related issue. Nevertheless, the behavioral aspects stem from upbringing. Early in life they learned to get their needs met in maladaptive ways, such as, being the bully. Although there are different types of difficult people, some are overly aggressive, while others may be passive-aggressive, but their dynamics are similar. Like all human beings, all they want is to be loved and accepted. Unfortunately, they have learned inappropriate ways to achieve this.

These behavioral patterns are deeply ingrained in the personality of the difficult person. The overly-aggressive difficult person (one who bullies, explodes, screams, etc.) uses their aggressive posture as a defense mechanism. Because of their weak and fragile ego, they need to protect themselves. Their best defense is a strong offense-aggression. Therefore, they feel in control of themselves only in a situation that allows them to feel powerful. But it doesn't stop there. Like all weak people, their insatiable need to feel secure makes it necessary for them to win and to win at any cost.

The second step in trying to cope with difficult people is to distinguish between a person who is having a bad day and one who is a difficult person. Keep in mind that difficult people make up a small percentage of the population. However, having an encounter with one makes that percentage appear larger. The first way to help distinguish between the two is to reflect on the history of the person. In other words, "Is the behavioral pattern normal or unusual for this person?" The difficult person is this way all of the time. A non-difficult person who is having a bad day is just reacting to a particular situation. Another approach in distinguishing between the difficult person and a person having a bad day is found in the way you communicate with them. Although hostile at first, the non-difficult person will eventually respond to your effective communication and rational reasoning. The difficult person will be relentless in their pursuit to beat you and win.

To help you maintain composure when confronted by difficult people, it is important to keep three things in mind.

- First, you can never change the difficult person. These people need to be this way and for them to change is to expose their vulnerability.
- When confronted by difficult people, remain focused and be firm. By assaulting your ego with insults and intimidation, they want you to lose control and fight with them. When this happens, they "got-ya." Listen to them, maintain direct eye contact and when appropriate speak in a clear firm

voice. It is easy to become wrapped up in the heated situation, so remain detached and distant from these people. Doing so helps keep you from becoming entangled in their web of misery and hostility.

- The final step that will help you cope with the difficult person is to not personalize the problem. Certainly, this is easier said than done. Between wishing they would be different, thinking you can really help them, and trying to survive their emotional assault, it's difficult not to internalize the problem. Yet, in order to cope effectively with these people, it is crucial to maintain your self-esteem.

Some of the following thoughts might be helpful in your attempt to depersonalize the situation:

- "This is their problem, I will not make it mine."
- "I'm not going to allow anyone to dictate my behavior."
- "They want me to fight with them, I won't allow it."
- "Their need to be difficult is a cover-up for their own inadequacies."
- "I have the choice to play or not to play this game."

The bottom line is that trying to cope with difficult people is never easy and is quite frustrating, especially when they can affect your service levels, project progress or overall support reputation. Trust the fact that all people have trouble dealing with difficult people. Although it may not seem possible to deal with difficult people effectively, remain confident in your abilities and coping skills. And keep in mind that engaging in an argument with these people is a no-win proposition. In fact, the only way for you to win is to elect not to play.

⚠

> *"The great enemy of clear language is insincerity."*
>
> *--George Orwell*

Communication Formats

Faxes, teleconferences, the World Wide Web, and other technological advancements guarantee that we can communicate with virtually anyone, anywhere. However, it is up to us to ensure that the messages we send are clearly understood by the recipient. Whether it is a face-to-face meeting or an overseas transmission, communication is a complex process that requires constant attention so that intended messages are sent and received.

Inadequate communication is the source of conflict and misunderstanding. It interferes with productivity and profitability. Virtually everyone in business has experienced times when they were frustrated because they just could not get through to someone. They felt as if they were speaking an unknown language or were on a different wave length. Communicating effectively is much more than just saying or writing the correct words. How we communicate is affected by frame of reference, emotional states, the situation, and preferred styles of communication.

There is a tendency for people to avoid unpleasant interactions, and sometimes managers will use written communication to avoid the discomfort of dealing face to face with staff. While written communication can play an important role in communicating about change, it should not be used for this reason alone. Below are some guidelines regarding the use of written, oral and electronic communication.

Oral communication is more appropriate when:
- Receiver is not very interested in getting the message. Oral communication provides more opportunities for getting and keeping interest and attention.
- Emotions are high. Oral communication provides chances for both you and the other person(s) to let off steam, cool down, and create a climate for understanding.
- You need feedback. It's easier to get feedback by observing body language and asking questions.

- The other person is too busy or preoccupied to read. Oral communication provides better opportunities to gain attention.
- You need to convince or persuade. Oral communication provides more flexibility, opportunity for emphasis, chances to listen to and remove resistance, and is more likely to affect people's attitudes.
- The details and issues are complicated, and cannot be well expressed on paper.

Written communication is appropriate if:
- You require a record of the communication for future reference.
- Your staff will be referring to details of the change later. You are communicating something with multiple parts or steps and where it is important that management understand them.
- Generally, it is wise to use both written and oral communication. The more emotional the issues, the more important it is to stress oral communication first. Written communication can be used as backup.

Electronic communication is appropriate if:
- You are required to provide periodic updates to upper management. Both email and Intranet provide an effective tool for global updates.
- You want to keep a very large number of users updated on systems events. This is especially important when the issue or updates are too technical for business management to proficiently relay to their staffs.
- You are communicating to an international business audience. It is difficult and can be quite expensive to bring global business leaders together, especially during on going communication. This is where email, Web and video conferencing become the most important tools for effective communication.
- As with written communication, it is generally wise to include oral communication along with any electronic. The more emotional the issues, the more important it is to stress oral communication first.

Oral and written (hard copy) communication used to be the dominant methods for overall communication. However, as the business industry becomes more and more global, electronic methods of communicating are now becoming more acceptable in our society. Unfortunately, these electronic methods are increasing the lack of physical interaction. Nevertheless, communicating with the business, whether verbally or electronically, is still a vital component to maintaining a strong, reputable systems department.

E-mail is now the most popular method of communicating. (And) users are becoming more and more creative with their formats. Here are some pointers to make your E-mail communication effective:

- *Keep it short.* One screen of information is preferable per message. If you need to say more, use an attachment.
- *Acknowledge all messages.* If someone goes to the trouble to write you a paragraph or two, at least tell them it arrived.
- *Sign all your messages in the body.* Include your E-mail address or phone number.
- *Avoid inappropriate cuteness.* But be more conversational than in traditional correspondence.
- *Lighthearted emoticons can help convey meaning.* Only if they are appropriate to your environment and your reader. However, because they are now so commonly used I am providing a list of the most popular meanings:

Symbol	Translation	Symbol	Translation	Symbol	Translation
: -)	smiley face/happy	8-)	eye-glasses	:-|	indifference
:-e	disappointment	:-P	wry smile	:-!	foot in mouth
:-&	tongue tied	;-)	wink	:-O	yell
:-/	perplexed	:->	devilish grin	:-Q	smoker
:-{	mustache	:-	male	:-(	frown/sad
:-@	scream	;-}	leer	:-D	shock or surprise
C=:-)	chef	d:-)	baseball smiley	>-	female

- When responding to a question or point, quote enough of the original message to help your reader remember what they asked or said.   Do not automatically duplicate their entire message each time. The E-mail message will then become too large to convey a clear message.
- Use an alias or handle for your personal E-mail account at home? When you work from home, be sure to change the

name that displays on your E-mail account for your business messages. You don't really want your boss to get a message from

Communication is your primary and most important tool for managing an efficient IT department. I have outlined some of the important parts of the business/technology communication process, but short of writing an entire book on the subject, it is difficult to discuss all the subtleties and issues about human communication.

There is no substitute for good judgement, and as an IT manager you need to be reflective and thoughtful about the ways you communicate. There is also no substitute for LISTENING, and receiving feedback from your staff and colleagues about how you communicate. You may make communication mistakes, but the mark of an effective leader is that these mistakes are quickly identified through feedback and discussion, and corrective action is taken.

In these times of doing more with less and increased use of technology, it's imperative to remember to do whatever you can to foster effective communication. Analyze and ensure that you are using the best possible method of communication for your industry. By looking at the world from another's point of view, your employees, co-workers, customers, and vendors will feel that you are really listening to them. Listening and responding in a way that makes sense to them will improve relationships, enhance performance, increase productivity, and positively impact the bottom line.

Business Networking
Business networking is the process of building strong, solid relationships that are mutually beneficial. These relationships involve people, and therefore there can be communication and other challenges that occur. Verbal communication styles differ. These differences can lead to some temporary stalls in the relationship.

The concept can be closely associated to technology networking, where in order to create an efficient technical communication platform you must ensure that all hardware and software aspects are compatible. However, different operating systems (languages) and different hardware requirements (switches/routers) make the interface process a complete challenge.

The challenges are the same in business communication. Most people you come in contact with will usually speak a different language. I don't mean people from different countries or cultures. I am referring to a situation where you and your contact are not communicating effectively. You know this is happening when you and your contact begin a conversation and before you know it, you do not understand what your contact means, or you feel you are not being understood yourself. If you add the e-mail factor- the lack of voice inflection, body language and eye contact, communication challenges become very real.

The problem lies in the perception and understanding of words and phrases, which can mean different things to different people. Also, some people use words freely without actually evaluating their meaning and ascertaining whether those meanings are exactly what was meant. Other people measure every word carefully, only using words with the exact meaning that is required. (And) cultural differences only magnify the problem.

The easiest way to work through these differences is to ask for confirmation. "If I understand correctly, this means..." is a good way to ask for confirmation. Also, asking for clarification by using choices will help make headway. Ask, "Did you mean this or that?" to help understand the contact's meaning.

The exact opposite problem is also very common- TOO much communication. In this situation, the relationship becomes a burden when one participant feels smothered or distracted by the frequency of the contacts. Although the best relationships are mutually-beneficial, in this situation, one participant benefits while the other feels burdened.

Evaluate the time requirements you impose on your networking relationships. Remember they have things to do besides help you. The idea of "frequency of contact" to help promote the relationship can backfire if that frequency is too much for one participant. If you are in a relationship where you feel smothered, speak up. The relationship will suffer if you don't. As the person requesting assistance, ensure that you are not imposing on your contact's time. No one is interested in helping someone who is too demanding.

In order to become a networking professional in business, you need to evaluate our verbal communication styles and requirements. Remember that most people are looking to achieve the same things. They may just have different methods for getting to the same place.

Web Based Communication

While the corporate IT group normally leads the charge to develop an Intranet (Internal + Internet), in the end it is up to communicators to provide for content. And make no mistake about it, content is what makes an Intranet a powerful communications tool. Once the pipeline has been built, however, it is your responsibility to ensure that only professional and ethical information flows through the pipes.

Intranets are not about having ``home pages'', although individual and departmental home pages are a natural evolution of the process. Intranets are about content; living, dynamic, fluid, timely, relevant and community. Communicators must make substantial contributions for an Intranet to succeed. This takes both vision and process.

Vision enables you to identify the many pockets of information that are available within your organization. Vision helps you to see that the most important goal of an Intranet is to facilitate work processes, and then to help make it a reality. Vision helps you understand that information will no longer be controlled from the top down.

With this in mind, communicators of the business become facilitators of *process*. An authoring process is needed to help people with access to the tools to be effective communicators. An approvals process is needed to ensure that top management, who are also learning during this process, are kept informed about what is happening. An acculturation process is needed, to introduce the new concepts and working styles that will evolve.

Business communicators can make a great contribution to their organizations by finding ways to make processes orderly and effective. However, any additions or changes requested by the business should be consistently analyzed to ensure that all content is adhering to corporate policy standards. Confidentiality and security are should be the primary goal. This type of administration can be a time consuming process for your team and in some cases, the process can become a full time job. Your goal as an IT manager is to ensure that there is a centralized and organized process for receiving and posting information.

Intranets can democratize information. Information is no longer the domain of top management and the communications department, and can no longer be controlled. As a result, your IT department should function more as a mediator to build consensus, to facilitate process, to moderate dialogues and to identify and reinforce credible sources of information.

Employees need value, substance and content, meaningful information that enables them to work smarter and be better at what they do. They need to learn to pull information from the system, instead of waiting to have it pushed at them from headquarters. Therefore, you must work with business communicators to develop new expectations and define solid goals.

Gauging your Audience

One of the most important aspects of communication, which is often taken for granted, is the use of jargon or technical terminology when communicating with the business. Many technical personnel assume that the recipient is technically savvy or knowledgeable enough to decipher the many acronyms and conversational concepts that exist in the technical world. The issue stems from years of study and on-hands experience. IT methodologies become so intense that the person seems to speak a whole different language, even during their personal off-duty lives.

IT today must bridge the gaps between management, your technology staff, and company employees. The first step is to identify the people in the organization who are genuinely excited about new technology. Learn enough about the people to know their level of technical expertise. Help them to see the business communicator's role as an integral one to the success of any type of Intranet project. Set up the pathways to information to ensure that those who need it will connect readily with those who have it.

During round table discussions, refrain from using extensive technical jargon. Help them feel comfortable during group discussions by allowing them to share input and feedback on the project. You may come to find that there are many valuable business contacts that can help make your job much less stressful.

Intranets are one of the most vital new business developments of this century. For the organization, an Intranet brings a new sense of community among employees, pride in the organization, new ways to work faster and smarter, and greater productivity. For business communicators, participating in an Intranet team will lead to career growth and personal satisfaction. Employees who embrace the Intranet and seek the communicator's role will become a crucial contact for your IT department.

Emotions in Communication

It only takes one person to create peace. The use of good communication skills will usually entice a reciprocal response. These skills can be used by either the high-power or low-power person. The most important thing to remember is to become aware of the way we communicate. Knowing how others perceive you will help you to respond to them or to clarify your intent. Observation is more than just observing what they say. Nonverbal communication clues can give you insights to what they are feeling even if they ca not or are not expressing it. Knowing your own prejudices should enable you to predetermine what you need to be aware of when you are in a situation that could prompt you to offend another.

Realize that emotions cause people to enter into an altered state. Their personal power has to alter its course, usually turning it to negative energy. To get to the heart of the conflict without escalating it, you need to express emotions and encourage others to feel safe in expressing their emotions. If someone expresses an emotion, you need to validate that emotion. Do not tell people that they "shouldn't feel that way." Emotions are real.

In our fast paced society, people often do not want to take the time to explain themselves or to listen closely to another. To listen actively you can do several things like nod your head, repeat important factors, or paraphrase their statements and ask if what you thought they said was what they really meant.

Listening to another person will give you clues as to how to solve your conflict so that you both the get what you really need. Seek first to understand, then to be understood. Listen for what they really want, and then use the information to order their environment so that both people can have their interests met. People can combine their positive energy to create an abundance of possibilities that would satisfy everyone's needs.

⚠

IT Management Lessons

- Work to align with the business. Build and maintain a strong relationship with all Sr. level business managers. The business will be more responsive to your project needs and be more understanding to issues that are out of your control.
- Make sure you regularly communicate the status and details of your ongoing projects to the business.
- Find the right method of communication for each business individual. Email is great for those who spend most of their time at their desk. For those who are frequent travelers, it is best that you create a hard copy memorandum.
- Gauge your audience carefully during meetings and conferences. Avoid using techy related jargon with less technical people.

Part VI: IT Customer Service

> *"Great things are not done by impulse, but by a series of small things brought together."*
>
> *--Vincent Van Gogh, Painter*

Aligning with the business

I have witnessed many IT mangers struggle to gain and maintain credibility with their customers. I hear frustration on both sides—IT managers frustrated at the pressure and lack of appreciation from their customers, and customers frustrated at the slow pace of systems implementation and poor quality of service. These disconnects are disappointing but often understandable, given that IT professionals are often unable to effectively manage their role of service providers by aligning their customers' expectations with their department's ability to deliver services. Here are some important lessons that can help align your IT department with your our own customers.

Attitude

Attitude is everything in business and in life, and there are hundreds of books available on how to build and maintain a success-oriented attitude. All these books basically relay the same message when it comes to customer service - managing your customers is essential to your success. However, there are a few basics that I feel require reiterating.

- IT professionals often disconnect with customers early by not having the right perspective. IT professionals consider themselves computer scientists rather than businesspeople or service providers. But IT is in the service business, not the science business. Your customers care as much about computer technology as automobile drivers care about combustion technology. It is your job to align yourself with your customer's needs and explain your services in *their* terms, not yours.
- The second attitude disconnect is the failure to understand the basic human psychology of expectation and satisfaction. Keep commitments. If you promise five widgets and deliver six, the customer is happy. If you promise five and deliver four, he/she is disappointed. They may get over their disappointment and adjust to the reality, but you will

not get full credit for your hard work. Resist the temptation to alleviate short-term stress by over-promising future products.

Planning

In managing expectations, planning means the art of anticipating your customer's next move. Falling behind on the latest business strategy(s) can end your career. Worst case scenario example, blindsided by the business when they announce that they are going to consolidate 12 call centers into 5 during the next 12 months, or they are doubling the size of the organization in the next 90 days. Amazingly, many IT managers have no idea where their business is headed or how to respond when it takes a sharp turn. This is a recipe for disaster, because the last thing you want is to be on the critical path to business success. Pressures and "assistance" from upper management can ruin even the best IT shop. How can you prevent these kinds of disasters?

Understand your business and your industry. Read the trade press and understand the forces that are acting on your industry. Listen to your internal customers, learn their plans and concerns, and find out what keeps them up at night. Unfortunately, many companies do not involve their IT departments as an element of their business planning. Non-integrated IT can become a bottleneck. Through planning, however, you can meet customer expectations and become strategically valuable to your business.

Let your customers know how you are planning to address their needs and what you will do for them. Emphasize the high quality of customer service that they will get from you. Do not think that you are being boastful: after all, if you keep quiet, your customers may not notice the difference. The technical and managerial skills you have acquired through years of experience can only surface at your command. Therefore, express and reiterate what you have to offer as often as necessary without over selling.

Physical Atmosphere

The physical environment of your technology workplace impacts two major factors of your employees' work life: their motivation and their productivity. A clean, comfortable, well-lit, and not overly noisy environment will go a long way toward motivating your staff to work more productively. The long-term results ultimately affect and improve customer service quality and support efficiencies.

First, give the following activity a try. Take a notepad and pen with you so you can make notes about your experience. Go outside the building and walk in, imagining that you are experiencing the place for the first time. Open your senses and take note of colors, sounds, smells, lighting, temperature, and anything else that strikes you. Walk into the data center and sit at a technician's workstation. How does the chair feel? What do you see? What do you hear? Put your hands on the keyboard. Are you comfortable? Open a file on the computer. Is it easy to read? Do your eyes feel strained? Do the monitor colors soothe, stimulate or stress you out? Take notes on every sense and feeling experienced. Look around at the walls and at anything else you see from the technician's workstation. Is there anything about these areas that either motivates or distracts you? Can you imagine a way to better utilize these areas?

The point is, there are all kinds of things in the physical environment (big, little, subtle and obvious) that affect employees' morale and productivity. Once you have assessed the physical environment of your workplace, the next step is to take measures to improve it. Following are some things you can do to optimize the physical environment for your employees:

- Encourage employees to bring in a few pictures or some artwork to personalize their work space. This will help them to feel comfortable at work and may even decrease their stress.
- Make sure employees have the equipment they need to work well. State-of-the-art technical equipment is nice, but it's not essential. What is essential is making sure your employees have the tools they need to do a good job. This includes tools, large monitors, special keyboards, phone systems, glare screens, wrist rests, and any necessary software.
- Locate computer monitors at least two feet from the eyes. If the screen can be read without strain, it is not too far away. Experts suggest that if the screen cannot be read, it is better to make the characters larger than it is to move the monitor closer to the eyes.
- Take measures to reduce noise. High ceilings, fabric-covered acoustic panels, and carpets help absorb noise. Installing sound masking and acoustic ceiling tiles can reduce noise by 40%. Wood furniture absorbs noise better than metal or glass. Doors to areas that are adjacent to the production floor (such as a power room or a room) should be kept closed.

- Pay attention to the lighting. Good lighting in a workplace has a positive effect on mood and productivity. It has been shown to decrease eyestrain, fatigue, computer errors, and even absenteeism.
- Make good use of walls, wallboards, and white boards. I suggest you use these areas to inform, motivate, encourage, educate, and whatever else you need to do on a daily basis. Think about it: These areas are in view of your technicians all day every day. It is a perfect opportunity to fill their heads with meaningful information!
- Last but not least, always maintain an organized and visually appealing work environment. Delegate daily cleanup responsibilities to each team member. I am not referring to throwing out the trash and cleaning the glass, which is facilities responsibility. I am referring to storing away tools and maintaining an organized troubleshooting environment.

The information above may seem insignificant to a fast paced technical support environment. However, the physical atmosphere is actually just as important as the technical atmosphere you provide for your employees. It produces calm emotions and a patient ambiance, which helps build a strong, understanding connection between your team and the business. To that end, it is safe to say that a comfortable employee is a productive employee.

⚠

> *"The most important thing in communication is to hear what isn't being said."*
>
> *--Peter F. Drucker*

Dealing with Angry Users

In the world of technology it is inevitable that you will have to deal with an angry user once, if not many times throughout your career. Angry and difficult users are a major cause of systems workplace stress, and they eat up huge amounts of your time and the resources of your organization. There are a lot of tricks and techniques you can use to deal with an angry user. However, I have found that it is more productive to focus on the most common mistake technician's make when dealing with the hostile, difficult or angry user. By avoiding this particular error, you can save yourself a lot of stress and time.

The number one mistake: When you are faced with an angry user, you or your team probably assume that the user wants his or her "problem" fixed. That is a logical approach and it is partly true. Angry users expect that you will be able to help them in some concrete way, by meeting their want or need. However, there is a small yet significant concept to keep in mind. Have you ever noticed that with a really angry person, even if you can "fix" the problem, the person still acts in angry or nasty ways? The reason is that angry customers actually want several things. Yes, they want the problem fixed, but they also want to be heard, to be listened to, and to have their upset and emotional state recognized and acknowledged.

What most technicians do with angry users is move immediately to solve the problem without giving that acknowledgment. The user is so angry that he or she is not prepared to work to solve the problem, doesn't listen, and gets in the way of solving the problem. So the number one error is moving to solve the problem before the customer is "ready", or calm enough to work with the technician. The result is the technician has to repeat things over and over (since the customer did not hear), and has to ask the same questions over and over. And that is what drives technology people nuts.

The Solution: The solution is to follow this general rule: When faced with an angry user, FIRST focus on acknowledging the feelings and upset of the user. Once the user starts to calm down as a result of having his or her feelings recognized, THEN move to solving the problem. You will find that this will save you and your staff a great deal of time and energy.

Here are a few example phrases you can use:
- *It seems like you're pretty upset about this and I don't blame you. Let's see what we can do about this issue.*
- *It has to be frustrating to have the printer jam every time you try to print.*
- *Most people would be angry if their file were lost or corrupt.*

Dealing with difficult user is a fact of technology life. These "negative" users can lead to employee burnout, low staff morale, or be responsible for someone leaving the industry entirely. However, if handled correctly, difficult users can be turned into some of the most loyal and helpful users you will ever see. Here are some basic steps to help you in turn user disconnects into opportunities.
- *Assume the user is telling the truth.* If you train your employees to always assume the user is truthful, you have just taken away a major source of stress related to technical support careers. The employee is not having a confrontation, nor are they conducting an interrogation. They are not looking for the negative, but listening to what is being told to them without having to be a judge that must rule in favor of the company because of a misplaced loyalty.
- *Let the user talk.* Let them air out the whole situation. This accomplishes two things. It allows the user to tell their story with all the details and emotion that they feel is necessary. This step is vital to let the user vent some of their emotion and anger. Do not say anything, except to give body language that you are listening intently. A good idea BEFORE you start the listening/information gathering process is to delegate all interruptions to someone else so that your entire attention is devoted to the user. If absolutely have to jump in and interrupt an anger or frustrated user, then interrupt their story with "Excuse me", "Just a minute", or "Pardon me" Otherwise, always listen without interruption or comments. Also remember that listening is the beginning of the information gathering

process for yourself, which is vital not only to rectifying the customer's problem, but to avoiding it in the future.

- *Be empathetic.* This is the step to (finally) begin communicating. Express understanding with how they feel or were treated. You are not admitting guilt. You do not even have to agree with them. You do, however, have to communicate understanding. Your tone of voice and body language both go a long way to reinforce what you are say. There is nothing worse then a manager coming over to a table with their hands on their hips, challenging, "is there a problem here?" In fact, without the proper tone or body language, your words will sound hollow.

- *Understanding.* This is the main step in reaching the user. This is where you ask any questions that you need in order to have the complete picture of the negative experience. Ask relevant questions to clarify your understanding of the facts. Resist jumping to conclusions until you are satisfied that you understand the entire situation.

- *Solution.* Solve the problem. Come to closure that you both feel good about. Remember the systems user was telling you the truth. Tell the user what you will do to rectify the situation. Make the user feel good about the solution. Do not sound angry yourself or make the user feel guilty. A good guideline is to deliver more than you promised. For example, if you said you would fix a printer issue, offer to enhance or modify their personal system environment by cleaning up temp files or creating application shortcuts. That not only solves the original problem, it brings them back again.

- *Follow-up.* If there is any way to follow-up with the user after the fact, you need to do it. Whether by e-mail, phone, or personally, this step is very impressive.

- *Take steps to fix the problem(s) that caused the problem in the first place.*
  A good idea to use help desk reporting software or keep a log of user issues which you can use to help you analyze trends and isolate redundant anomalies. Remember that the best IT managers prevent problems rather than just fix them!

The bottom line to any technical support approach is to make sure you address the feelings first, THEN move to fix the problem. Listen, talk quietly, validate, soothe. Once your customer has calmed down, you can resolve the problem and thus the root emotion. When you do this, the user becomes bonded to you in ways you will not believe. Their confidence in you triples, their trust in you solidifies, and your future with them is secure. Furthermore, they will sing your praises forever. All it takes is a little understanding and you can turn an angry client into a happy user forever.

> *"Obstacles are those things you see when you take your eyes of the goal".*
>
> *- Unknown*

The "No" Circumstance

No matter what language you say it in, customers do not like to hear the word *no*. Regardless of whether you are in Mexico, Germany, France, or Japan, your customers want what they want, when they want it, and how they want it. When customers do not get what they want, the result is usually disappointment, frustration, and upset.

As IT manager, the responsibility of saying *no* will usually fall in your lap at the end of the day. Playing liaison under the *no* circumstance can actually be one of the most difficult tasks of this position. The challenge is due to the fact that many companies believe that service is *giving customers what they want*, and of course, when you can you should. However, the problem is that IT support, especially desktop technicians, feel committed to providing exactly what the customer wants. In other words, they have an awkward time saying no. (And) when they do they feel helpless and often fail to use techniques that may bring about a satisfactory conclusion to a difficult situation. When it comes to saying *no* three facts are important to keep in mind:

- Sometimes circumstances force you to say *no*.
- Saying yes does not guarantee a happy customer.
- Saying *no* does not mean that you have to end up with an unhappy customer.

Like it or not, circumstances exist that require you to say no to your customers. These include the following examples:

- *Federal regulations.* Certain rules and regulation may be imposed on the business (depending on the industry you are in) by an outside government agency. For example, if you work for a financial institution, you would have to say *no* to a customer who requests changes to the infrastructure in order to accommodate more users. FCC regulations prohibit the modification of the LAN infrastructure and would require an audit and revalidation of the communications environment. The process can be costly and time consuming.

- *The law.* Companies must comply with federal and state laws. For example, some states allow Internet liquor sales and advertising, where others consider it a crime to sell alcoholic beverages online. As a technology manager it is imperative that you familiarize yourself with the local state laws prior to planning your e-commerce strategies.
- *Company policies and procedures.* Each and every organization will have their own set of policies that everyone in the organization, including the technology department, must adhere to. System security is probably one of the most important aspects in avoiding breach of policies. For example, confidential information related to Human Resources, quality departments, financial departments and Customer databases rely heavily on the security of your networks and servers.
- *Out of stock.* Although your aim is to maintain a 10 percent inventory level of replacement parts or supplies, there may be occasions where items may be unavailable by the vendor. Products such as PBX interface cards, network switch components, and even printers may be on back order by the supplier.
- *Just not possible.* You will often find that users will ask for applications or desktop computers that are just not available in your environment. Most requests stem from new software and new computers that they may have at home. Users would love to replicate their work environment to match the latest and greatest available to the public.

Given the examples above only justifies the need, but the fact remains that you still have to say *no*. There are two different ways for doing this:

- *The hard no:* Finding the right way to say *no* is always difficult, but even more difficult is saying *no* to a customer who insists on getting their way no matter how justifiable the *no* may be. In cases such as this, you may try to offer alternative solutions to the issue. However, keep in mind that this approach can quickly turn into a catch 22 if you are not fully prepared with good alternative solutions. Therefore, as long as you are certain that you cannot provide the service or product to the customer you must flat out state *no* and stick to your guns. This will obviously bring disappointment to the customer, but as long as you avoid treating your customer like an adversary, they will have to respect and accept the response. Furthermore, avoid one-liner responses in a monotone voice. This type of response reflects an "I don't care attitude" that can quickly create a tense situation.

Examples of one-liners include:

That's not our policy.
That's not my job.
I'm not allowed to do that.
I have no idea.

Examples of body language that accompany these responses:

A blank stare
Head held down
Look-away eyes
Distracted fidgeting

- *The service no:* Giving a customer a hard clear *no* is always uncomfortable and should be avoided if possible. However, there is no intermediate way of saying *yes* or *no*, you either can or cannot provide the request. Wording your responses in a way that the *no* sounds like a *yes* will only cause unrealistic expectations for the customer. The unclear answer may also be interpreted as a lie and may lead to a distrusting relationship between you and the business. The alternative answer to a hard *no* is a service *no*. Providing a service *no* usually requires having alternative solutions to the request. Fortunately, most situations can be resolved using this approach, along with two simple phrases.

These phrases are:

What I will do is...
The phrase tells customers that you want to help them, along with the specific actions you will take to get their problem(s) resolved. The alternative actions you offer may not be exactly what the customer wants but will usually help create an acceptable resolution to the problem and reduce the customer's feelings of frustration.
What you can do is....

The second phrase tells customers that they have some control over the outcome of the situation and that you consider them your partners in getting the problem resolved. Possible suggestions for customers may involve recommendations for a temporary fix to the problem or actions that the customer can take in the future to prevent this occurrence from happening again.

Here is an example using the two phrases:

> *Situations one:* A user calls and states that they have lost some vital files that were located on a network drive. The user also accuses someone in his department of accessing the private secure folder and deleting the files.
> *Solution:* After sincerely expressing your concern over the loss of the files, explain how you plan to address the issue. "*What I can do is* review the backup tapes for the files. If I find them I will notify you immediately. *What you can do is* send me your login ID and the date of the last time you logged in to the drive. This will help me determine whether you or someone else deliberately removed you files.

Bottom line, the key to gaining respect from a customer, whether you say *yes* or *no*, is always asking yourself the question: "What does this customer need, and how can I provide it to the best of my ability?"

Δ

> *"You must look within for value, but must look beyond for perspective."*
>
> *--Denis Waitley*

International Technical Support

The definition of the ideal IT environment is one that is designed to exceed the enterprise's strategic goals while nurturing the individual to achieve exceptional productivity and job satisfaction. However, with international business fast becoming a norm in every industry, IT managers are faced with finding the right balance between local and cross-cultural support differences. In other words, effectively mentoring and managing customer service skills that can stretch over to any type of cross-cultural business environment.

There are four key competencies that you need to keep in mind in order to successfully apply your customer service skills across borders:

- *Personal influence:* being aware of what is happening interpersonally and being able to use a wide repertoire of skills to make things happen. There is a real danger that your level of comfortable influences may not be effective enough to make things happen.
- *Managing ambiguity:* business today is uncertain and complex with fewer clear-cut answers. The ability to tolerate and work within these grey areas becomes very important. Stay sharp and keep an open mind.
- *Making decisions:* at the end of the day IT managers need to reach conclusions and move forward. This can be much harder when working internationally. Managers have to transfer knowledge into action, exercise discretion to make sound decisions and assume responsibility for making things happen.
- *Customer focus:* it is easy to get wrapped up in the difficulties of managing across borders and the customer can get lost in the process. Customers' needs must be continually assessed as an integral part of the decision making process and solutions developed to meet their needs.

My experience with many international managers suggests that these competencies are applicable in a wider context. For example, managers working across national boundaries need to be able to read cultural signals successfully and to act on them appropriately.

Unfortunately, the two do not always go hand in hand. If you are not very good at reading cultural signals, but you get on with it anyway, you are likely to make all kinds of mistakes. Almost as bad is where you read the signals and fail to act on them, becoming in effect a dispassionate observer. This is a mistake in an international service context, where pro-activity is key. What you want to do is both read the signals and act on them appropriately.

International IT managers need to take account of all the stakeholders in the support delivery chain. You need to be able to respond to the diverse needs of colleagues, business partners, employees, contractors and suppliers, as well as the end user.

Reading the context the user is operating in is crucial. This includes not just the country, but the region, the industry they work in, the organization they work for, the sort of technology they are used to handling, how they view their profession and their role, maybe even their gender. During my time in Mexico, I found that every American manager that arrived to work in a multi-cultural environment experienced a steep learning curve. They often made incorrect assumptions about employees, which were all based on American culture traits. Needless to say, the mistakes they made along the way often affected the outcome of many tasks and projects.

Many of the differences are centered on a small number of key dimensions, for example:

- *Different attitudes to power and status.* For example, a Mexican manager will quickly stop progress on a project or meting if he/she feels that there is insufficient respect by American colleagues. Issues similar to this have been known to cost a company thousands of dollars in project setbacks.
- *Different rules to operate within.* For example, French managers prefer to know precise objectives and procedures and Mexico managers seem to be happier to act and work out the how's and why's later. An important example of individual versus group. Individuality is very strong in the US. In contrast, Mexico managers value group decisions and consensus.
- *Different values.* During development programs with a group of Mexico managers the importance of relationships surfaced time and again, more so than with many English and US groups. Hence, the Mexican culture looks to family values as a basis for doing business.
- *Different attitudes to expressing emotions.* In some countries blunt discussions are not acceptable, even in a business

context; in others it demonstrates a willingness to explore new ideas without restraint. Similarly, the US and northern European countries view of time is sequential, where one event leads to another and can be planned and budgeted. The southern European approach is to juggle a series of tasks, so taking a phone call in a meeting is perfectly acceptable. In contrast, answering a phone during a meeting with Mexico counterparts is considered impolite.

To avoid problems, think about your own preferred communication style. Does the organization have a common style? How much are national styles influencing you and the customer? You may have to adapt the way you communicate when you are dealing internationally. Be aware of the cultural context. Think carefully before you speak about the likely impact of your words; you may need to be more formal and careful in the use of humor. Be wary of stereotypes; they may be a useful template but they conceal as much as they reveal. At best they are a starting point for further exploration; at worst they are totally misleading.

For short business assignments or visits, briefing on points concerning national characteristics which affect conducting business can be useful. However, it can take years to break down cultural barriers. Training and development to promote international team-working and consciously communicating successes in cross-cultural working can play an important part.

Additional International Focus

Always be conscientious of your surroundings when traveling. Analyze the current situation at hand and forget about any previous trips or conversations with other cultures. Make it a habit of aiming for the future and ignoring the legacy of the past. Applying this concept will help you build a stronger future by drawing on your current diversity strengths. In addition try applying the following steps for success:

- *Ask yourself some hard questions.* Do you have key influencers who champion the international customer? Is recruitment based on genuinely international criteria? Do you have a clear understanding of the diverse needs of your international customers? Do you talk the customer's language, not just in words, but in a genuine understanding of what is important to them? Workshops can provide a rich understanding of the culture and reveal an agenda for change. For example, during a series of workshops, a technical support services business concluded that its culture was based on an unshakeable belief that it would endure and that mistakes would be punished. These attitudes led to poor customer service, an unhelpful risk aversion and heavy, standardized controls. In order to be successful, the company found that it had to move to a more flexible international culture format of support.

- *Set improvement goals.* An essential part of successfully managing change is to involve others. Individuals need to understand and be committed to make personal change. An international workforce is typically well educated and interested in the business. This demands clear and regular communication of the process.

- *Implement these goals.* What leaders do to support the international customer is undoubtedly very powerful and can be helped along by symbolic actions. When I worked at Providian Financial I helped to promote the removal of separate offices for managers, to encourage a more open, flexible and less hierarchical customer support environment. The strong feelings this generated suggested the true nature of this symbolic act! Training and development can be a significant vehicle of cultural change.

- *Maintain the momentum.* Building an international focus is not a one-off event. People leave and new people join and the customer profile changes. A continuous process of reinforcing, updating and sustaining is required.

The most important thing is to learn from mistakes. In order to work successfully across cultures, you will need to adopt management approaches that take account of the complex differences between colleagues and customers. Start with a candid assessment of yourself and your organization, including assumptions and stereotypes held. Valuing alternative points of view and demonstrating respect for other people are crucial skills, but hard to make a reality. Misunderstandings will inevitably arise. How you handle them and learn from them will determine whether you make a success of working and managing internationally.

▲

> *"Life's most persistent and urgent question is: What are you doing for others?"*
>
> *--Dr. Martin Luther King Jr.*

Improving Customer Service

One of the best ways you can provide and improve your technical support services is to identify key customers and negotiate reasonable expectations. Once those are agreed upon and consistently met, your IT customer service can truly be optimized.

Consistently meeting or exceeding the expectations of your key customers is one of the best ways to ensure that you are providing excellent customer service. For example, you may design a change-management process that dozens or even hundreds of technical engineers may be required to use. But you would probably only need to involve a handful of key technical representatives to assist in designing the process and measuring its effectiveness.

- *Meet their expectations.* Customers who expect more than they receive will be dissatisfied, no matter how good the service is. If your department has a reputation of giving customers techy talk or future promises to keep them at bay, that is a recipe for disaster.
- *Believe their complaints, not their vision.* Customers are not necessarily the best determiners of technology choices. They tend to define their needs based on what they know, rather than what is possible. However, if you offer options, customers can select among the options and articulate flaws. Customers are best able to help correct flaws in interfaces, usability issues, and functional deficiencies. These complaints should not be criticized, but noted and corrected as soon as possible.
- *Empower your customers.* A common complaint I hear from IT is that customers set up their own "shadow IT" departments and do not use the institutional systems. This is particularly true in high-tech organizations with an abundance of computer savvy staff, who are often frustrated at the slow pace of change and the lack of control over systems they use daily. There are two solutions to this problem. (1) Is to become an IT dictator, refusing to allow your customers the freedom to move forward on their own.

This is may not always be a good solution, as your customers will revolt. (2) The second solution is to empower the customers by giving them some control over setting priorities.

- *Involve your customers.* Customer involvement is essential for the success of any system. I still meet IT managers who believe that they can work with the customers through a requirements-definition process, create a design document, and then turn it over to their developers for delivery. Customers must be involved at each step, through techniques such as functional walkthroughs, conference room pilots and beta testing, and frequent discussions about business plans and future needs. IT must assume that every system will change throughout its lifecycle.

- *Avoid asking technical questions.* Many technical managers tend to over involve the customer by allowing them input to highly technical questions such as the number of servers needed to manage their environment. Answers to technical questions should remain the responsibility of the IT manager.

- *Measure the quality of your service through feedback.* Provide feedback forms to your customers on a regular basis. Independently verifiable performance metrics are essential in satisfying the customer. A good IT department measures all aspects of their performance and regularly communicates the quality of their work to their staff. There are several important impacts of metrics. First, everyone knows the quality of the work, and often can use the metrics to prevent problems from becoming critical. Second, metrics become the basis of objective discussions with customers about the acceptability of service, and the cost of making it better. Customers will not support your efforts to improve service unless you can objectively demonstrate what they receive and why.

- *Keep your attitude positive and your frustrations in check.* IT management is not primarily a technical function. It is a service business. When you become frustrated with your customers, remember that you are in your job because they have theirs. You and your staff should remain focused on where the revenues if coming from. I see frustrated IT managers who wish for a better class of customers. Take your customers as you find them and work with each

situation. If you become frustrated and lose your poise, you
will lose your ability to communicate, and "push back"
instead of listening. Your customers will become
dissatisfied with your service. Maintaining a positive
attitude is the key to customer satisfaction, as well as other
things in life.

Remember: the customer (i.e. user, manager, or vendor) is the
most important part of your organization! They are the ones that can
make or break your technology team. Without a high percentage of
satisfied customers you are not going to be leading for very long. In
fact, you can have the greatest technical team in the world, but without
the greatest customer service your team will never be fully successful.

⚠

IT Management Lessons

- Maintain a comfortable atmosphere for your employees.
  Provide them with the right tools and literature to help them get
  the job done. You will find that your staff will have a better
  attitude towards work. An attitude that will stretch over to their
  customer service approach.
- Train your staff to be patient and understanding when dealing
  with angry users.
- If you absolutely have to say 'No' to a special request, make
  sure you have sufficient data to justify your response.
- Focus on the cultural aspects of your environment when
  working on technical issues over the border or abroad.
- Obtain customer feedback and work to improve any complaints.

Part VII - Managing your Environment

> "Failing organizations are usually over-managed and under-led.
>
> --Warren G. Bennis

Managing IT Services

One way to create admiration for your IT organization is to provide high-quality IT services. Today's customers have little tolerance for downtime, slow performance, and slow responses to requests. Thus, with rising customer expectations, you need to provide high-quality services today and increasingly higher levels of service in the future.

Service management should be the cornerstone of your IT organization; covering everything related to planning, providing, monitoring, coordinating, and reporting service quality. Service management drives all aspects of managing an effective IT organization. Through a structured service level agreement, service management establishes the basis for IT service delivery processes such as availability, capacity, continuity, security, and production operations management, as well as support processes such as problem management, configuration management, and release and change management (among others).

The following steps and processes should help you get started and moving in the right direction:

Define Your IT Services

It is important that you begin this process by identifying and describing the services that your IT organization provides. Describe these services from the customer's perspective, covering service levels and main service components. The main objective is to completely understand your IT organization's services.

Focusing on IT services is a key element to aligning IT with desired business outcomes. Some examples of IT services include hosting services (floor space, rack space, uninterruptible power, network connectivity, and bandwidth), server configuration and management services, streaming media services, email services, and application management.

Define Your IT Service Levels

With well-defined service levels, the next step is describing and quantifying expectations for the services that your IT organization provides to its customers. You also need to define service levels that your vendors provide to your IT organization. The main objective is to set specific targets to evaluate performance and provide a foundation to define formal agreements for the quality of service. Without appropriate performance measurements, your IT organization may not be able to provide acceptable levels of service. For example, IT service levels should cover items such as these:

- Hours of operation (such as 24x7 support)
- Connectivity (redundant network connections)
- Availability (99.99 percent uptime)
- Capacity
- Accessibility
- Backup and restore
- Disaster recovery
- Call response
- Problem management

Identify IT Growth Projections

To plan capacity, you need to quantify the expected growth in IT service demand. The best projections are based on models utilizing built-in feedback mechanisms to ensure both accurate inputs and planning assumptions. IT growth projections may also include strategies for meeting anticipated growth. The main objective is to understand IT performance and capacity characteristics in order to determine their impact on IT services. With an understanding of the factors that affect the demand for IT services, you will be able to determine the right capacity to achieve desired service levels and optimize the use of IT resources.

IT growth projections cover a range of metrics, including the growth in the number and/or utilization of various aspects of the IT infrastructure. This may include applications, servers, users, transactions, networks, staff, space, and environmentals. For example, server projections may be based on planned growth rates, square feet per cabinet, servers per cabinet, and cabinets per square feet of data center space. Network projections may include bandwidth projections based on sales forecasts for new customers and servers, and bandwidth for growing existing customers.

Identify Critical IT Assets

Identify the key components of your IT services, covering people, process, and technology. Without these components, the committed service levels cannot be achieved. The objective is to understand critical IT assets and their impact on IT services. With an understanding of critical IT assets, you can make special provisions for availability and security to achieve committed levels of service. For example, critical assets may include servers and storage systems for business-critical applications and data. They may also include backbone network components.

Identify IT Opportunities and Risks

IT opportunities are conditions that could be exploited to improve IT services for the benefit of your business. Conversely, risks are threats and vulnerabilities that may need to be mitigated. The objective is to understand your IT organization's strengths, opportunities, weaknesses, and threats so that appropriate actions can be taken. At any point in time, you need to be aware of opportunities for and threats to your IT organization's ability to achieve committed service levels. Some examples are SWOT (strengths, weaknesses, opportunities, and threats) analysis; risk assessment; business impact analysis; and security assessments.

Establish IT Control and Design Objectives

Establish key criteria for the design and management of IT services and underlying IT infrastructure. The main objective is to identify and understand the criteria to assess, build, and manage IT services, and provide linkages with IT services and the underlying elements of the IT infrastructure. With well-defined control and design objectives, you can achieve service levels in a cost-effective way. The following sets of principles and criteria are examples of controls:

- CoBIT control objectives
- SysTrust principles
- ITIL assessment areas

Design objectives may include high levels of any or all of the following:

- Availability
- System performance
- Systems and data integrity
- Scalability
- Security

- Manageability
- Interoperability
- Accountability
- Integrity
- Confidentiality
- Auditability

Your IT organization's success in providing high-quality IT services depends on its ability to align IT services and the underlying IT infrastructure. Structure and organize your IT architectures and processes to ensure that availability, performance, and security are aligned with IT service commitments, and do it cost-effectively. Furthermore, outline plans of action to help you build and manage the key technologies and processes. This will enable your IT organization to achieve its desired levels of service.

Λ

> *"If everything seems under control, you're not going fast enough."*
>
> *-- Mario Andretti*

Risk Management

Risk management is the process in which potential risks to your environment and the business are identified, analyzed and mitigated, along with the process of balancing the cost of protecting the company against a risk vs. the cost of exposure to that risk. Business is risky. And with the increasing dependency of businesses on technology to maintain and advance their organizations, the risks and the stakes are greater than ever. Therefore, as IT manager you play a major role in identifying and mitigating risk issues.

The question is how should you manage these risks? Effective risk management is a multi-step process. Having a really thorough understanding of what you have is the most important step (e.g. technology, applications, processes, security, etc.). In a lot of cases, the exposures that come up and bite you are because of things you were not aware of. In order to mitigate risk, you must take time to understanding what all the different pieces are and how they all fit together.

For instance, your company may be planning to build a data center, but from a technology standpoint there is a 90% chance that the project will not completed on time. You need to look at the various costs associated with mitigating that risk. You may choose to allocate or spend more money to pay for contractors who can assist you in getting the job done faster. Or, if the risk of completing the project late is deemed too great, your company may decide to push back the deadline. That decision would force the business to estimate potential lost revenue or productivity losses, as well as calculate the costs associated with extending the deadline for the project.

Self-awareness is a crucial. You have to recognize what the business is trying to do. You need to be able to look at that broad picture. I like to think of it as looking at the trees and the forest at the same time (not to be confused with Active directory). However, being self-aware is only the first step in effective risk management. Another essential component is planning for possible failures. It may sound simple, but in the course of operating and maintaining a technology environment, it is an often overlooked task.

Amazingly, there are still many IT manager's out there who do not effectively plan for inevitable failures. They do not take the proper

steps in understanding what it is going to take to recover from a failure
(e.g. disaster recovery processes). As a support entity, you should
assume that lack of a (recovery) process in <u>any</u> area of your
environment could generate losses to the business. Hence, every aspect
of your technology environment, from applications to hardware, should
be considered a business critical function.

Minimizing Downtime

From an information technology perspective, risk management
includes minimizing an organization's exposure to downtime or loss of
service from its IT systems or processes. From a business process
standpoint, risk management is more about managing a "portfolio of
systems and projects" in order to maximize financial returns on those
investments and minimize the potential for conflicts and delays. Once
a you and the business have recognized what the potential risks are, it is
equally important to evaluate how costly those risks can be, as well as
how much time and money should be invested in mitigating those risks.
That process, known as business impact analysis, is another crucial
component of effective risk management for companies.

A business impact analysis really helps define what a company's
losses would be. If you were to have a power outage, even as short as
15 minutes, what are your financial impacts, what are your non-
financial impacts, how are your customers affected, how is your
department image affected?"

Cost Considerations

Once you have determined what the risks are and what their
losses might be, you must then decide whether or not to address each
risk. To do so, consider the size of the risk and its consequences to the
organization.

You might choose to accept greater risk of failure because there is
greater reward. For example, a mail-order business evaluates the risks
of launching a Web site in time for the holidays. There may be many
risks involved with the project, including the possibility that the Web
site might not generate adequate sales volume and may result in a loss
on the project investment. However, the potential rewards of operating
an online business during the busy holiday shopping season might be
great enough for the company to decide to go forward.

If, on the other hand, a particular risk is relatively unlikely but the
potential cost to the company is great, then the organization might
choose to address the issue in advance. For example, an Arizona-based
IT service provider is unlikely to suffer power outages due to

hurricanes or earthquakes. But since the company's financial losses or liability resulting from a power outage could be significant, it might decide to install a backup power system to protect itself.

In the end, failing to address risk management is perhaps the greatest risk of all for any IT manager. What you may wind up doing is fighting a lot of fires. As with any business, time is always of the essence, and it is not uncommon to see the time consciousness IT person shortchange a lot of the planning. Surprisingly, many IT leaders wait until they get to that bridge before addressing the concept. Unfortunately, not recognizing that there are multiple bridges and some of them are already falling down.

Λ

> *"An optimist is someone who goes after Moby Dick in a rowboat and takes the tartar sauce with him."*
>
> *--Zig Ziglar*

Budget Basics

One of the biggest challenges for an IT manager is to develop a budget that accurately reflects not only the initial cost of technology equipment, but all related expenses as well. To make certain that your environment remains an efficient tool for your company, a basic understanding of the technical and economic realities is necessary.

When it comes to (IT), the issue for many organizations is one of understanding and expectations. For example, companies tend to treat desktop computer purchases as a one-time expense. In other words, they buy it and then forget about it. Any technology person knows this is obviously not a cost-effective measure for the purchase of a desktop system. Computer systems require maintenance and support. (And) with the ever-increasing rate of change in technology, the state-of-the-art machine today will only be cutting-edge for a few months, and close to obsolescence in two to three years.

Keeping up with technology is undeniably expensive, but there are strategies to minimize expenses and prevent surprises from decimating your budget. You should consider the Total Cost of Ownership (TCO) when purchasing any type of systems related equipment. Using the desktop computer again as an example, only 30 percent of the total cost of owning a desktop is the initial purchase of hardware, software and peripherals. Seventy percent of the ownership cost goes to technical support, repairs, training and upgrades. As systems get larger and include networks, email, Internet access and more complex databases, the yearly cost for just one computer can run close to $10,000 (when you include the salaries of technical support people and lost productivity due to hardware failures.) Therefore, if a computer system costs $2,000, be aware that maintaining the system will likely cost you $7,000 or more.

Creating the budget

As an IT manager, I feel the most stressful (administrative) time of the year has to be budget process and approval time. However, it does not have to be as painful if you follow a few guidelines.

- *Write a five-year plan and update it annually.* Base your budget on a plan that rationalizes the numbers and looks forward to the future. In the IT industry we do not really know what is going to happen in five years, but it is important to have goals. As the forecasts get closer to the year being budgeted, technologies and budgets will come into sharper focus. Include a budget forecast section that you revise annually. This forecast will help management allocate resources over a longer planning period.

- *Know what you need vs. what you would like to have.* Include in your budget at least a small portion of what you would like to have but that you do not consider necessary. Lay it out in terms of what the extras will buy. For instance, increased productivity, better information, or faster information. Do not expect to get it all. Draw a distinct line between what you believe is absolutely necessary and what you think will be helpful. Then, be quite direct about the risks and negative consequences of not giving you what is required to get the job done. For example, you might request additional bandwidth in response to user complaints about Web site latency. Since the Web site is an increasingly critical component in getting your organization's message out, be firm in your request for the allocation.

- *Include a budget review.* It is important to relate how you used the money you asked for last time around. Doing so demonstrates to management that you know what you are doing. Many times, budgets will never work out as expected, often for reasons beyond your control. For example, a sudden and unexpected shift in technologies can render a project obsolete. Included a history with the current request will give you a chance to explain why certain things did go as planned last year.

- *Divide the budget into projects, infrastructure, and operational expenses.* Operational expenses are the day-to-day expenses for maintaining the current infrastructure. Be sure to present operational expenses separate from the rest of the budget, and definitely list them as a necessity. Projects and infrastructure, meanwhile, are the "value-added" component of your department. Projects and infrastructure are your department's contribution to the overall goals and objectives of the organization. Divide them into projects, which add value to a specific department, and infrastructure, which benefits the whole organization.

- *Research costs carefully.* When creating budgets, always conduct careful line-item research on projected purchases of equipment, software, and services. Include prices as they stand unless you are positive that a new development will reduce the price significantly. Then, during the year, increase the value by buying at a higher technology level or increasing the number of items purchased rather than leaving a large surplus at the end of the year.

- *Hit your targets.* If significant and unexpected price/technology changes happen, for instance, if the price of RAM doubles or halves, document these changes thoroughly and explain them in your budget review. Budgets are designed to present the possibilities. Careful planning will help you to hit your numbers. Management will gain confidence in you if you can demonstrate consistent performance.

- *Include tutorials in your plan.* Business Sr. management may not understand why you need to upgrade to Layer 3 switches throughout the network. Provide enough information to give them a sense of the capabilities they are buying by including a short and basic paragraph about the business advantages. They will then feel empowered to make the decision to fund it. Diagrams and charts help make difficult concepts concrete. Above all, make it short and sweet. Sr. managers have too much on their plates to hear the full history of the switch change and the impact it will have on the organization.

- *Write the plan.* Start with a review of the past year. Then present your goals (a single statement of purpose) and objectives (specifics needed to reach each of the goals). Present your updated five-year forecast. Describe details of the plan with a discussion of tasks, timelines, and required resources in a project detail section. Finally, include an itemized cost section. Management will be able to communicate the things they consider important in your next year's plan by either suggesting projects you had not considered or (unfortunately) slashing projects that may be close to your heart.

- *Sample budget outline*

1. Review of past year: Highlight your successes and changes to last year's plan.
2. Goals and objectives: Give a succinct overview of the basis for your plan.
3. Forecast: Make a rolling five-year forecast of your plans, including costs, in bullets.
4. Plan details: Outline key operational areas and specific projects.
5. Costs: Give realistic costs for each project and operational expenses.

With these methods and a year or two of practice, you will be able to get the financial resources you need to do your job. You will also earn the reputation as someone who communicates well, has good organizational skills and has a strong understanding of technology as a business resource.

⚠

> *"Resolve to be a master of change rather than a victim of change."*
>
> *--Brian Tracy "*

Managing Through Change

There is no other field that experiences more regular and complex change than the Information Technology arena. There is no way around it, change is inevitable and anyone not willing to adapt will be phased out of the field as quickly as they arrived. I used to joke by saying - "If I were a Doctor and I studied and tested as much as I do as a systems tech, I would be a Neurosurgeon by now". Troubleshooting is very similar to the medical field; diagnose and treat. The difference is a human body and its functions never change. Doctor's demonstrate their knowledge through yearly refresher testing and they upgrade their techniques through regular courses in the areas of new treatments and drugs.

In contrast, an IT technician or engineer must re-certify his or her knowledge on new technologies and procedures almost daily. As I mentioned earlier in the readings, the IT field has changed so dramatically that old school technicians sometimes have a difficult time finding their place in the field.

Unfortunately, the IT workload, like the universe, is expanding at an ever-accelerating rate. The customer continues to require more secure, privatized, customized and economized 24/7 support, and fulfilling that consumes an immense amount of time and labor. When change is imposed (as in downsizing scenarios), clearly the most important determinant of "getting through the storm", is the ability of leadership to lead.

If you are to manage change effectively, you need to be aware that there are three distinct time zones where leadership is important. They are *preparing for the journey, Sailing through the storm, and destination arrival.*

In an organization where there is faith in the abilities of formal leaders, employees will look towards the leaders for a number of things. During drastic change times, employees will expect effective and sensible planning, confident and effective decision-making, and regular, complete communication that is timely. Also during these times of change, employees will perceive leadership as supportive, concerned and committed to their welfare, while at the same time recognizing that tough decisions need to be made. A climate of trust between leader and the rest of the team must existence in order to bring hope for better times in the future. That makes coping with drastic change much easier.

In IT organizations characterized by poor leadership, employees expect nothing positive. In a climate of distrust, employees learn that leaders will act in indecipherable ways and in ways that do not seem to be in anyone's best interests. Poor leadership means an absence of hope, which, if allowed to go on for too long, results in an organization becoming completely nonfunctioning.

As an IT manager you must deal with the practical impact of unpleasant change, but more importantly, you will struggle with employees who have given up, have no faith in the system or in the ability of your leadership to turn the organization around.

Preparing for the journey

It would be a mistake to assume that preparing for the journey takes place only after the destination has been defined or chosen. Preparing for the change journey means leading in a way that lays the foundation or groundwork for ANY changes that may occur in the future. Preparing is about building resources, by building healthy teams from the start. Much like healthy people, who are better able to cope with infection or disease than unhealthy people, technology organizations that are healthy in the first place are better able to deal with change.

As a leader you need to establish credibility and a track record of effective decision making, so that there is trust in your ability to figure out what is necessary to bring your organization through.

Sailing through the storm

Leaders play a critical role during change implementation, the period from the announcement of change through the implementation of the change. During this middle period the organization is the most unstable, characterized by confusion, fear, loss of direction, reduced productivity, and lack of clarity about direction and mandate. It can be a period of emotionalism, with employees grieving for what is lost, and initially unable to look to the future.

During this period, as an effective leader you need to focus on two things. First, the feelings and confusion of employees must be acknowledged and validated. Second, you must work with your staff and begin creating a new vision of the altered workplace, helping them to understand the direction of the future. Focusing only on feelings may result in wallowing. That is why it is necessary to begin the movement into the new ways or situations. Focusing only on the new vision may result in the perception that the leader is out of touch, cold and uncaring. A key part of leadership in this phase is knowing when to focus on the pain, and when to focus on building and moving into the future.

Destination arrival

In a sense you never completely arrive, but the initial instability of massive change has been reduced. People have become less emotional, and more stable, and with effective leadership during the previous phases, are now more open to locking in to the new directions, mandate and ways of doing things.

This is an ideal time to introduce positive new change, such as assessment of control procedures or Total Quality Management. The critical thing here is that you must now offer hope that the organization is working towards being better, by solving problems and improving the quality of work life. While the new vision of the organization may have begun while people were sailing through the storm, this is the time to complete the process, and make sure that people buy into it, and understand their roles in this new organization.

Leadership before, during and after change implementation is THE KEY to getting through the storm. Unfortunately, if you have not established a track record of effective leadership, by the time you have to deal with difficult changes, it may be too late. Playing a leadership role in the three phases is not easy. Not only do you have a responsibility to lead, but as an employee yourself, you have to deal with your own reactions to the change, and your role in it. However, if you are ineffective in leading change, you will bear a very heavy personal load. Since you are accountable for the performance of your staff, you will have to deal with the ongoing loss of productivity that can result from poorly managed change, not to mention the potential impact on your own enjoyment of your job.

Changes Driven by the Business

One meaning of managing change refers to the making of changes in a planned and managed or systematic fashion. The aim is to more effectively implement new methods and systems in an ongoing organization. Many times the changes to be managed will lie within and will be controlled by the business organization. However, these internal changes might have been triggered by technical events originating outside the organization, in what is usually termed "the environment." Hence, the second meaning of managing change, namely, the response to changes over which the organization exercises little or no control (e.g., technical advances, the actions of competitors, and shifting economic tides). Managing the kinds of changes encountered in many of the larger organizations requires an unusually broad and finely-honed set of skills. The main skills required include the following:

- *Political Skills.* Organizations are first and foremost social systems. Without people there can be no organization. Lose sight of this fact and any change leader will definitely lose his or her head. Organizations are intensely political. (And) the lower the stakes, the more intense the politics. As a change leader you should dare not join in this game, but you had better understand it. This is one area where you must make your own judgments and keep your own counsel; no one can do it for you.

- *Analytical Skills.* Make no mistake about it, as a change leader you had better be very good at something, and that something better be analysis. Fortunately, IT requires a significant amount of daily analytical skills. The key is to apply the same analytical troubleshooting skills to change. Two particular sets of skills are very important here: workflow operations or systems analysis, and financial analysis. You must learn to take apart and reassemble operations and systems in novel ways, and then determine the financial and political impacts of what they have done. Conversely, they must be able to start with some financial measure or indicator or goal, and make their way quickly to those operations and systems that, if reconfigured a certain way, would have the desired financial impact. Those who master these two techniques have learned a trade that will be in demand for the foreseeable future. This trade, by the way, is called "Solution Engineering."

- *People Skills.* As stated earlier, people are the heart of the organization. Moreover, they come characterized by all manner of sizes, shapes, colors, intelligence and ability levels, gender, sexual preferences, national origins, first and second languages, religious beliefs, attitudes toward life and work, personalities, and priorities. These are just a few of the dimensions along which people vary. And as an IT leader you should be prepared to deal with them all. The skills most needed in this area are those that typically fall under the category of communication or interpersonal skills. To be effective, we must be able to listen and listen actively, to restate, to reflect, to clarify without interrogating, to draw out the speaker, to lead or channel a discussion, to plant ideas, and to develop them. All these and more are needed. Not all of us will have to learn Russian, French, or Spanish, but most of us will have to learn to speak Systems, Marketing, Manufacturing, Finance, Personnel, Legal, and a host of other organizational dialects. More important, we have to learn to see things through the eyes of these other inhabitants of the organizational world. A situation viewed from a marketing perspective is an entirely different situation when seen through the eyes of an IT person. Part of your job as a change leader is to reconcile and resolve the conflict between and among disparate points of view. Charm is great if you have it. Courtesy is even better. A well-paid compliment can buy gratitude. A sincere "Thank you" can earn respect.

- *System Skills.* There is much more to this than learning about technology environments (i.e. computers, switches, routers, etc.). Most people in today's work routine do need to learn about computer-based information systems. However, from a business perspective an information system is an arrangement of resources and routines intended to produce specified results. To organize is to arrange. A system reflects organization and, by the same token, an organization is a system. Hence, you need to be careful when communicating with business management. You need to understand their perspective of a system. There are "hard" systems, which of course is the physical layout of your domain, and there are "soft" systems: compensation systems, appraisal systems, promotion systems, and reward and incentive systems.

- *Business Skills.* Simply put, you had better understand most of the business functions in your organization. In particular, you need to understand how the business departments interface and how the overall business process of your environment works. This entails an understanding of money and where it comes from, where it goes, how to get it, and how to keep it. It requires you hone into your knowledge of markets and marketing, products and product development, customers, sales, selling, buying, hiring, firing, EEO, AAP, and just about anything else you might think of.

Organizations are highly specialized systems and there are many different schemes for grouping and classifying them. Some are in the retail business, others are in manufacturing and distribution. Some are profit-oriented and some are not for profit. Some are in the public sector and some are in the private sector. Some are members of the financial services industry (i.e. banking, insurance, and brokerage houses). Others belong to the automobile industry, where they can be classified as original equipment manufacturers (OEM). Some belong to the health care industry, as providers or insurers. Many are regulated, some are not. Some face stiff competition, some do not. Some are foreign-owned and some are foreign-based. Some are corporations, some are partnerships, and some are sole proprietorships. Some are publicly held and some are privately held. Some have been around a long time and some are newcomers. Some have been built up over the years while others have been pieced together through mergers and acquisitions. The point to all this is that no two businesses are ever exactly alike.

Understanding the skill sets required to managing change is only a format of adaptability. That is, changes and the change problems they present are problems of adaptation to an ever-changing set of circumstances. The key is to know how your people will respond. The following strategies should help your approach:

Strategy	Description
Rational-Empirical	People are rational and will follow their self-interest once it is revealed to them. Change is based on the communication of information and the proffering of incentives.
Normative-Reeducative	People are social beings and will adhere to cultural norms and values. Change is based on redefining and reinterpreting existing norms and values, and developing commitments to new ones.
Power-Coercive	People are basically compliant and will generally do what they are told or can be made to do. Change is based on the exercise of authority and the imposition of sanctions.
Environmental-Adaptive	People oppose loss and disruption but they adapt readily to new circumstances. Change is based on building a new organization and gradually transferring people from the old one to the new one.

Generally speaking, there is no single effective change strategy. It is all based on your type of business and/or technology environment. However, a proven approach is to adopt a mix of different strategies that you can assess and choose from. Furthermore, there is no perfect way of managing your change. That is, you should manage it pretty much the same way you would manage anything else of a turbulent, chaotic nature. Thus, it is more a matter of leadership ability than management skill.

Λ

> *"Without data, all you are is just another person with an opinion."*
>
> *--Unknown*

Inventory Administration
Software Asset Management
The Creativity, independence and ambition that drive many corporations to success can also drive IT departments to distraction. If you have a variety of qualified people throughout your enterprise purchasing and installing software and hardware, you are going to have a hard time keeping track of all your IT assets. Add in staff turnover and it is nearly impossible. For example, how many laptops, desktops, servers and software packages do you have? What are the expiration dates for commonly used applications? Do you have too many or too few software licenses?

Organization is the key. No fair using an outdated Excel spreadsheet or sending an intern with a clipboard to each department. Asset management software is the way to go. It helps control many points of entry for purchasing hardware and software assets. It will help avoid being a victim of your own environment. Thus, you need to carefully manage your software to ensure a return on the company investment.

Software asset management can help identify and avoid any outdated software on your equipment, identify whether you have the correct licenses for the software held and establish if you have the proper technology to meet your goals. Managing your software and keeping legal can be a challenge. Using illegal or under-licensed software puts your company at risk and if you want reliable, productive systems your software must be legal and you have to be sure that it is.

The real way to protect your business is good software purchasing and management. It may be the case that your business has just grown too fast and the proper procedures for purchasing and licensing software have not been put in place. As a first step, think of putting one of your staff in charge of software, someone with responsibility for purchase, distribution and control of where the software is kept. Also, an asset database should be kept, which keeps track of the software you have, and the status of the corresponding licenses.

Good software management generally consists of two steps: performing a software inventory, and establishing maintenance processes. This involves analyzing your installed software products and comparing them with the licenses you own. Depending on the size of your business and the condition of your records, this may be quite simple. Scan all computers' hard drives in your environment and summarize the software installed. For example if you are using a Microsoft platform and products, you can use Microsoft Software Inventory Analyzer to create an inventory report listing the core Microsoft products on your equipment. Otherwise try one of the popular commercially available for your type of environment (i.e. Novell, Unix, Linux, etc.).

Keep track of the software installed on each desktop system. An inventory tool can provide the following:

- *Summarization of the results.* Providing all the information necessary to easily manage your software environment.
- *Efficient documenting of your licenses.* Providing easy reference to your licenses or proof of purchase documentation, summarize the total licenses owned and establish any licensing excesses or deficiencies. Helps save time in preparation for internal or external audits.
- *Ongoing management:* Provides assistance with administration, management and ongoing maintenance of your software inventory. Helps discover if anyone is using outdated or illegal software. Thus, maintaining environment standardization by upgrading existing licenses on a timely basis. This in turn leads to greater productivity among employees who will spend less time dealing with file sharing and transfer issues.

Consider centralizing your software acquisition processes by clearly identifying a software administrator who is responsible for ensuring that the policy is understood, implemented and followed. Other items to consider include: Who currently makes the acquisition decision and based on what criteria? Do you buy your software from one source? If you have a license shortfall, focus on volume licensing.

Depending on the size of your company, you should create a set of basic policies and procedures. This is an important step in reducing liability for copyright infringement. Create a written company policy regarding the acquisition and storage of software. Most importantly, communicate the policies clearly to all users and staff who buy, manage, or use software in your organization.

Once you have your software inventory, it is important to keep it up to date. Your policy can state who is responsible for the software, who records the serial and registration information and what happens to redundant software. Also, consider whether you permit personally owned software to be installed and under which circumstances. Include all users in the process. Ask them what software programs might make their jobs easier or more efficient. Taking steps to get organized and included the user community. This can help increase business productivity, not to mention employee job satisfaction.

Hardware Asset Management

For many organizations, "managing hardware" simply means managing to gather enough money together to buy it. However, while planning hardware purchases is definitely crucial for technology management, there are some other important considerations if you are going to maintain your investments and avoid technology mistakes. Keeping track of your technology gives you the information you need to plan for the future, avoids outdated/depreciation trouble, and saves you from misspending your money.

The main eye opener came with the Y2K scare. Many organizations were forced to confront a long neglected issue, inventory. As the media exploited the panic, many organizations found themselves scrambling to show that they were prepared to deal with a potential major disaster. As the years go by it is hard to remember how serious this seemed back then, especially with the fact that so little actually happened.

Many organizations did not know how many computers they had, let alone the configurations and software that resided on them. Substantial amounts of time and money had to be invested to figure out where threats to mission-critical operations existed and how to prevent those threats from becoming a major problem. This was possibly the only positive outcome of the Y2K scare. It forced IT leaders to examine how they were managing (or mismanaging) technology and where they could improve. The situation is also forced many managers to start a technology inventory and asset management process that may otherwise have been ignored.

There are several other reasons to maintaining an inventory of your technology assets, besides preparing for the end of a millennium:
- *Save Money.* If your organization is considering an upgrade, you need to know what your current computer configurations are. If you do not have enough RAM on half of your computers for a new operating system, you will waste a lot of money buying licenses for all of your computers. And you will probably need the serial numbers or license certificates to qualify for the often substantial upgrade discounts.

- *Find Problematic Systems.* Tracking your computers and the problems associated with them, can give you vital information about which technology is working in your organization, and which is not. Finding out that older desktops crash on a daily basis due to memory or processor issues could help justify your proposal to purchase new ones.
- *Avoid Hardware Warranty Headaches.* If you find that some of your computers or servers were purchased over five years ago, they most likely are out of maintenance warranty. An updated inventory can help you keep track of your hardware prevent costly repairs that would other wise be covered by a manufacturers warranty.

Now that you are convinced about the need for maintaining a technology inventory at your organization, some direction is needed. Depending on the number of computers and the complexity of the systems within your organization, there are a number of ways to keep track of everything.

- *Do It Yourself (DIY).* For those of you who do not have many computers, sometimes the simplest solution is best. Creating an inventory database, spreadsheet, or paper inventory form may give you the custom solution you need, without all the complications some software packages present. DIY databases can easily be posted to your web page for administrative reference.
- *Use Asset Management Software.* For larger organizations or networked offices, DIY solutions may not work for you. Instead, you may need to use some type of software to keep track of what you have. Many times this software can be configured to run over a network and keep track of your computers and software automatically. Take some time to research the pros and cons of some of the most current applications available.
- *Maintaining your Inventory.* A common problem in many organizations is losing data accuracy and hard copy documentation. Establishing a central repository where only one administrator can update the data is your best solution. CD's, manuals, and documentation should be kept in one safe location. If you are managing a large organization, the inventory backups should be treated as regular data backups and stored off-site on a regular basis.

Bottom line, asset management is not an enjoyable taks. I would be lying if I said I enjoyed it. However, in the long-run it can make your IT management life easier and more productive. A good inventory of your technology hardware can help your organization avoid wasting money, track problems with your current configuration, and keep everything nicely organized.

Δ

IT Management Lessons

- Make sure monitor and maintain a consistent, acceptable SLA percentage.
- Identify and manage risks before they become full scale issues.
- Learn to accept, modify and role with change. Send your team the message that change can be a good thing.
- Research and prepare before creating a finalized budget.
- Maintain a well organized and detailed asset inventory database. The data can help you prepare for audits. It can also be a major time saver in trying to identify and locate devices (e.g. through MAC addresses, asset tags or IP's) during troubleshooting.

Part VIII: Project Management

> *"A wise man learns from his mistakes, a wiser man learns from the mistakes of others"*
>
> *-- Unknown*

Take Charge

From small application projects to large infrastructure re-engineering, it is my opinion that many IT managers rarely take into consideration the amount of project management (PM) responsibility that awaits them. I have also found, through years of project experience, that many IT managers make "on the fly" decisions that usually end up in failed project tasks and late deliverable timelines. Both domestic and international IT manager's must, on occasion, change mental gears and focus on projects as a full time project manager would.

All IT managers, regardless of their exact management role in technology, need to address the characteristics of project management in an organized manner. Preparation is the key to succeeding and adapting to the project world. (And) believe me, if you plan to expand your management duties into the International sector, you will definitely require these skill sets. Why? Because in my broad global experience and as cross-cultural business environments demonstrate, resources are limited. You may not have the luxury of having a dedicated project manager available, even for the most complex and large-scale project.

There is a great deal of information available that can help you become a better project manager. However, unless you plan to make project management a full time career, the data available can be over kill for an IT leader. Therefore, I have compiled a set of project management success points, characteristics, and approaches that I feel will help you survive and succeed with your project goals. The following are crucial elements that will help you create a successful game plan:

- *Focus on three dimensions of project success.* Simply put, project success means completing all project deliverables on time, within budget, and to a level of quality that is acceptable to the business and stakeholders. As IT manager you must keep the team's attention focused on achieving these broad goals during each and every project.

- *Planning is everything.* The single most important activity that project managers engage in is planning. Detailed, systematic, team-involved plans are the only foundation for project success. And when real-world events conspire to change the plan, you must make a new one to reflect the changes. So planning and re-planning must be a way of life as you manage projects.

- *Feel and transmit to your team members a sense of urgency.* Because projects are finite endeavors with limited time, money, and other resources available, they must be kept moving toward completion. Since most team members have lots of other priorities, it is up to you to keep their attention on project deliverables and deadlines. Regular status checks, meetings, and reminders are essential.

- *Use a time-tested, proven project life cycle.* We know what works. Models such as the standard ISD model and others concepts can help ensure that professional standards and best practices are built into your project plans. Applying these models not only help support quality, but they help to minimize rework. So when time or budget pressures seem to encourage taking short cuts, it is up to you to research and identify the best project life cycle for the job.

- *All project deliverables and all project activities must be visualized and communicated in vivid detail.* In short, you and your team must early on create a tangible picture of the finished deliverables in the minds of everyone involved so that all effort is focused in the same direction. Avoid vague descriptions at all costs; spell it out, picture it, prototype it, and make sure everyone agrees to it.

- *Deliverables must evolve gradually.* It simply costs too much and risks too much time spent in rework to jump in with both feet and begin building all project deliverables. Build a little at a time, obtain incremental reviews and approvals, and maintain a controlled evolution.

- *Projects should require clear approvals and sign-off by the business.* Clear approval points, accompanied by formal sign-off by business upper management should be demarcation points in the evolution of project deliverables. It is this simple: anyone who has the power to reject or to demand revision of deliverables after they are complete must be required to examine and approve them as they are being built.

- *Thorough analyze the need for project deliverables.* Research has shown that when a project results in deliverables that are designed to meet a thoroughly documented need, then there is a greater likelihood of project success. So managers should insist that there is a documented business need for the project before they agree to consume organizational resources in completing it.

- *You must fight for time to do things right.* Projects must have available enough time to "do it right the first time." Therefore, in the event of unforeseen project delays or issues, you must fight for this time by demonstrating to top business managers why it is necessary and how time spent will result in quality deliverables.

- *Responsibility must be matched by equivalent authority.* It is not enough to be held responsible for project outcomes; you must ask for and obtain enough authority to execute these responsibilities. Specifically, you must have the authority to acquire and coordinate resources, request and receive SME cooperation, and make appropriate, binding decisions which have an impact on the success of the project.

- *Business management must be active participants, not passive customers.* Most project sponsors rightfully demand the authority to approve project deliverables, either wholly or in part. Along with this authority comes the responsibility to be an active participant in the early stages of the project (helping to define deliverables), to complete reviews of interim deliverables in a timely fashion (keeping the project moving), and to help expedite access to subject matter experts (SME), members of the target audience, and essential documentation.

- *Projects typically must be sold, and resold.* There are times when you must function as a salesperson to maintain the commitment of the business. With project plans in hand, you may need to periodically remind people about the business need that is being met and that their contributions are essential to help meet this need.

- *You should acquire the best people you can and then do whatever it takes to keep the nonsense out of their way.* By acquiring the most skilled, the most experienced, and the best qualified people, you can often compensate for too little time or money or other project constraints. You should serve as an advocate for these valuable team members, helping to protect them from outside interruptions and helping them acquire the tools and working conditions necessary to apply their talents.

- *You must actively set priorities.* In today's leaner, self-managing organizations, it is not uncommon for technical project team members to be expected to play active roles on many project teams at the same time. Ultimately, there comes a time when resources are stretched to their limits and there are simply too many projects to be completed successfully. In response, some organizations have established a Project Office comprised of top managers from all departments to act as a clearinghouse for projects and project requests. The Project Office reviews the organization's overall mission and strategies, establishes criteria for project selection and funding, monitors resource workloads, and determines which projects are of high enough priority to be approved. In this way top management provides the leadership necessary to prevent multi-project log jams.

Defining Team Member Roles

Your project team typically includes people from different parts of the organization, with different skill sets and operating styles. You may not have worked extensively with these people before. Your project usually has a tight time schedule, and your team members most likely are working on several other projects at the same time.

Success in this environment requires that you reach agreements about how you will work with your team members and staff to maximize everyone's contribution and minimize wasted time and mistakes. You need to develop an approach that gives you and others confidence that everyone will live up to their commitments. You need to understand the planned roles, and you need to be comfortable with them.

The following concepts can help you define and clarify how team members should relate to each other and to their assigned tasks:

- *Authority:* The ability to make binding decisions about your project's products, schedule, resources, and activities. Examples include your ability to sign purchase orders not to exceed $10,000 and your ability to change a scheduled date by no more than two weeks.

- *Responsibility:* The commitment to achieve specific results. An example is your promise to have a draft report ready by a set date.

- *Accountability:* Bringing consequences to bear based on people's performance. Having your boss reflect in your annual performance appraisal that you solved a difficult manufacturing problem is an example of accountability. Many people think of accountability as a negative concept (if you drop the ball, you pay the price). This fear often causes people to shun positions in which they would be held accountable for performance. Paying a price when you foul up is certainly half of the concept. The other half, however, is that when you do a good job, you are rewarded. This positive reinforcement is a far more effective way to encourage high-quality results. I totally believe that we should always give credit where credit is due.

These three terms address similar issues. However, each one is a distinct element that is required to define and reinforce team relationships. In other words, we must distinguish the similarities and differences of the elements. Consider authority and responsibility, as follows:

- *Similarity:* Both authority and responsibility are upfront agreements. Before you start your project, you agree who can make which decisions and who will ensure that particular results are achieved.

- *Difference:* Authority focuses on process, while responsibility focuses on outcomes. Authority defines the decisions you can make but does not mention the results you have to achieve. Responsibility addresses the results you will accomplish, with no mention of the decisions you can make to reach your desired outcomes.

Consider responsibility and accountability, as follows:

- *Similarity:* Both responsibility and accountability focus on results.

- *Difference:* Responsibility is a before-the-fact agreement, while accountability is an after-the-fact process. People who make promises, fail to keep their promises, and experience no resulting consequences, create some of the worst frustrations in a project environment.

It is essential that you keep in mind the guidelines for accountability. If you are responsible, you should be held accountable. In other words, if you make a promise, you should always face consequences based on how well you honor your promise. If you are not responsible, you should not be held accountable. If something goes wrong but you were not responsible for ensuring that it was handled correctly, you should not face negative consequences. Of course, you should not receive positive accolades if things go well in this case, either. Holding people accountable when they are not responsible is called *scapegoating*. This process of assigning blame to the closest person when things go wrong only encourages people to avoid dealing with you in the future.

Moving forward, the next step is to assign project roles. This is where your delegation skills become a crucial tool for creating a successful project team. To delegate is to give away something you have. You delegate for three reasons: To free yourself up to do other tasks, to have the most qualified person make decisions, and to develop another person's ability to handle additional assignments prudently and successfully.

Determining what you can and cannot delegate during project task assignments requires that you follow two main guidelines:
1. You can delegate authority, but you cannot delegate responsibility and;
2. You can share responsibility.

You can choose to transfer to another person the right to make decisions that you are empowered to make, but you can not rid yourself of the responsibility for the results of those decisions. For example, suppose you have the authority to sign purchase orders for your project not to exceed $10,000. Suppose further that you are not told not to give this authority to someone else, and no policy specifically prevents you from giving it to someone else. You could delegate some or all of this authority to your lead tech if you wanted to. That is, you could give your lead the authority to sign purchase orders for your project not to exceed $10,000. However, if your lead mistakenly bought 50 cases of Cat5 cable for $3,000 instead of the 30 cases that were actually required, you would be responsible for the poor decision. Thus, you can always take back authority that you delegated to someone else, but you can not blame the person for exercising that authority while he or she has it.

▲

> *"Before I built a wall I'd ask to know what I was walling in or walling out."*
>
> *-- Robert Frost*

Project Effectiveness

As many IT professionals say: "Turn every job into a project, it improves workflow and gives us opportunities to build positive relationships." From a project perspective, your success depends on your team working enthusiastically together toward completing the project goals. To understand more about the concept, you need to recognize that projects have unique characteristics and are different from operational type work.

The following items are unique characteristics of projects and IT management:

- *Something must be done which has not been done before.* There is something unique or different about each of your projects. It might be time allotted or the budget. You might be working in a different geographic area or with different team members. No matter how much your projects are alike, they are also unique in some way. You must keep your eyes and mind open so they discover those unique aspects as early as possible. If you do not recognize the uniqueness of each project, you cannot respond effectively.
- *The project ends with a specific accomplishment.* All projects must have a predetermined goal and objective. We want to produce a product, build an infrastructure, deliver a software program to the business or improve a process or system. At the end of each project you can look at the tangible results of your work. This is one of the best features of project leadership. You and your team can look back and say, "Look what we accomplished."
- *There is a beginning, an end, and a schedule for completion.* By definition, projects have a finite life. It may be days, weeks or years but there is an anticipated completion date. Utilizing project management tools and techniques, you will see that good time estimates and scheduling are critical for project success.
- *There is a budget. Resources are limited.* Few, if any, of us ever get a blank check for our projects. There are always

limits to the money available, people who can work on the project and other resources. You must first determine what resources are available and then effectively integrate and allocate those limited resources for project success.

- *Other people are involved in your project on an ad hoc basis.* Projects are completed by teams. Other employees, bosses, contractors, outside consultants, clients or customers may all be involved at some point. Communicating and coordinating with these people are essential skills of a successful IT manager. Effective communication cannot be overemphasized for team success. People who know what is going on always out produce people who do not.
- *Phases and activities are sequenced.* If you build a house, you have to put in the foundation before installing the walls and the roof ca not be built until the walls go up. On every project, regardless of the type, there is an ideal sequence of events that effectively uses the project resources to meet the project goals. This sequence of events balances the triple constraints of time, cost and performance standards.

Identifying the characteristics of your projects is just the first step in effective management of your projects. The next step is identifying the goals and projected outcome of the project. This process may seem simple, but many IT managers rarely have a clear, documented version of the project vision. Most managers that I have known seem to feel that a virtual vision of the project (i.e. adjust and adapt as you go) is just as productive. However, the outcomes of their projects were usually quite different than expected.

The following phases are critical to identifying your detailed goals:

- *Define* - Clearly defining your project is critical. What outcome or accomplishment do you expect to have at the end of your project? What goal or goals do you want to achieve? What is your strategy to meet your goals? Defining your project and its goals is the first step in successful project management. Projects are and should be very specific. A poorly defined project may end up accomplishing the wrong goal.
- *Plan* - Planning is the most often overlooked step in project management. Projects that end poorly can usually be traced back to poor planning. As the old axiom goes, "He didn't plan to fail, he just failed to plan." Good planning does not expend time, it saves time.

- *Implement and Control* - Implementation is the fun part of project management. This is where all the action happens. During the implementation time frame, you have to keep the project team motivated and involved in the project. Effective monitoring and project control is crucial for success. The sooner problems, delays, changes or other situations are known, the better you can address them, adjust course and meet your project goals. There are various ways to monitor and control projects. Thus, take some time to research the different types of processes and/or tools available.
- *Completion* - By definition, projects have to come to an end. Remember, projects have a beginning and an end. Project completion or wrap-up is an important part of project management. Have you ever been on a project that would not end? Endless projects can end up being failed projects. Project wrap is used to review what happened, what went well and what did not. No project is over until the paperwork is complete. There may be reports required, audits to conduct and other paperwork to be completed. Use the project completion phase to learn from your projects so that the next one will be even more successful. The "wrap up" is a good time to review team members' performance and to thank them for their efforts and support. Do not forget to thank your people for a job well done!

Effective and efficient project management is important in every aspect of technology. It is the combination of art and science. Using the steps and phases detailed above, you will improve your chances of successfully completing projects on time and within budget.

Project Communication

The key to successful projects is effective communication. That is, sharing the right messages with the right people in a timely manner. Informative communication creates continued buy-in and support from key audiences and team members, prompt and responsive problem identification and decision-making, a clear project focus, ongoing recognition of project achievements, and productive working relationships among team members.

Planning your project communications upfront enables you to choose the appropriate media for sharing different messages. Project communications can be both formal and informal. *Formal* communications are planned. They are conducted in a standard format in accordance with an established schedule. *Informal* communications occur as people think of information they want to share. Informal communications occur continuously in the normal course of business. However, you must take care not to rely on these informal interchanges to share thoughts about all aspects of your project, because they often tend to involve only a small number of the people who may benefit from the topics being addressed.

To minimize the chances for misunderstandings and hurt feelings, try the following:

- Confirm in writing the important information that you share in informal discussions.
- Avoid having an informal discussion with only some of the people who are involved in the topic being addressed.

Written reports enable you to present factual data more efficiently, choose your words to minimize misunderstandings, provide a historical record of the information you shared, and share the same message with a wide audience. However, written reports do not do the following:

- Allow your intended audience to ask questions to clarify the content, meaning, and implication of the message you are trying to send.
- Enable you to verify that your audiences received and interpreted your message in the way you intended.
- Enable you to pick up nonverbal signals that suggest your audience's reactions to the message.
- Support interactive discussion and brainstorming about your message.

Most important, you may never know whether your audience even read the reports you sent! Take the following steps to improve the chances that people will read and understand your written reports:

- Prepare regularly scheduled reports in a standard format. Doing so makes it easier for your audience to know where to look for specific types of information.
- Stay focused. Preparing several short reports to address different topics is better than combining several topics into one long report.
- Minimize the use of technical jargon and acronyms.

- Use written reports to share facts, and identify a person for people to contact to clarify or discuss further any information included in those reports.
- Clearly describe any actions you want people to take based on the information in the report.
- Use novel approaches to emphasize key information, such as printing key sections in a different color or on colored paper, or mentioning particularly relevant or important sections in a cover memo.

After you send your report, discuss with the people who received it one or two key points that you addressed in the report. Doing so will tell you quickly whether they have read it. Keep your reports to one page if possible; if not, include a short summary (one page or less) at the beginning of the report.

Λ

> *"A meeting is a gathering of important people who singly can do nothing,*
> *but together can decide that nothing can be done."*
> *--Fred Allen*

Meetings

For today's high-technology companies, solutions to problems are not so simple that one individual can provide all the answers. A broad variety of skills, knowledge, and background is needed to address most issues. The result is meetings to define problems, generate solutions, develop strategies, and on and on. (And) few words can elicit the emotional reactions of anger and frustration than the word meeting can. People view meetings as being anything from an indication of non-interpersonal contact to the biggest time-waster in business today.

Companies devote great amounts of valuable time to meetings. Yet, employees often describe meetings as wasted time debates that produce tenuous solutions not supported by key employees or social hours where little is accomplished. Afterwards, people may be confused about what was decided or who is responsible for following through.

Here are some more of the most common frustrations about meetings that I hear from people:

- Not being given sufficient notice
- Not having the right people attend
- Not starting on time
- Not having an agenda
- Not sticking to the agenda if one exists
- Having no actions result from the meeting
- Discussing issues that you thought had been resolved at a previous meeting
- Having people represent that they have the authority to make a decision and then having their decision reversed after the meeting
- Reading written material aloud that people could have read themselves beforehand
- Having 95 percent of the meeting deal with issues in which you are not interested or involved
- Knowing that the next meeting will be no better than the last one

Meetings can be very valuable when they are planned and managed effectively. When handled correctly, meetings can help you learn about other team members' backgrounds, experience, and styles; stimulate brainstorming, problem analysis, and decision-making; and provide a forum for people to explore the reasons for and interpretations of a message.

You can improve your meetings by taking the following steps.

Pre-meeting

- Clarify the purpose of the meeting. Build a clear, concise agenda.

- Decide who needs to attend and why. If you need information, decide who has it. If you want to make decisions at the meeting, decide who has the necessary authority and make sure that that person attends.
- Give plenty of notice about the meeting.
- Tell others the purpose of the meeting.
- Prepare a written agenda that includes topics and times. Doing so helps people to see why attending is in their interests. The agenda is also your guideline for conducting the meeting.
- Circulate the written agenda and any background material in advance so that people can prepare for the meeting.
- Keep meetings to one hour or less. You can force people to sit in a room for hours, but you cannot force them to keep their minds on the activities and information discussed. If necessary, schedule several meetings to discuss complex issues or multiple topics.

Meeting conduct

- Start on time, even if people are absent. When people see that you will wait for latecomers before starting, everyone will come late!
- Assign a "timekeeper". Someone who will remind the group when allotted times for topics have been exceeded.
- Take written minutes of who attended, what items were discussed, and what decisions and assignments were made.
- Keep a list of action items to be explored further after the meeting, and assign responsibility for all entries on that list.

- If you don't have the right information or the right people in attendance to resolve an issue, stop your discussion and put it on the action item list.
- End on time.

Follow-up
- Promptly distribute meeting minutes to all attendees.
- Monitor the status of all action items to be performed.
- Do not just think about these suggestions; act on them!

Every meeting should conclude with a summary of work completed, a clear action plan for outstanding tasks, and a decision about subsequent meetings. The summary should relate directly to the purpose: What was the goal? Was it achieved? What remains to be done?

The action plan should list specific tasks, the person(s) responsible, and the completion date for each. Resolve any confusion and adjust the plan as needed. Get the next meeting on the schedule while everyone is present. Check with participants in a few days to make sure they can complete follow-up tasks. The time you spend in preparation and follow-up will pay off with meetings that begin and end well.

△

> *"Life is full of obstacle illusions"*
>
> *-- Grant Frazier*

Managing Cross-Cultural Projects

Differences in approaches, values and expectations between customers, suppliers and team members with different cultural backgrounds have lead to many project failures. By understanding the impact of cross-cultural differences, IT leaders can increase the probability of an international venture's success—from setting up a new plant over the border to providing top quality systems support to non-American clients.

Miscommunication across cultural lines is usually the most important cause of cross-cultural problems in multinational projects. Miscommunication can have several sources, including:

- *Differences in body language or gestures.* The same gesture can have different meanings in different parts of the world. For example, Bulgarians shake their heads up and down to mean no. In addition, the way people count on their fingers is not universal: The Chinese count from one to ten on one hand, and eight is displayed by extending the thumb and the finger next to it. The same gesture is interpreted as meaning two in France and as pointing a gun in North America.
- *Different meanings for the same word.* Like gestures, words can have different meanings or connotations in different parts of the world. The French word "char" means Army tank in France and car in Quebec. The word "exciting" has different connotations in British English and in North American English. While North American executives talk about "exciting challenges" repeatedly, British executives use this word to describe only children's activities (children do exciting things in England, not executives).
- *Different assumptions made in the same situation.* The same event can be interpreted many different ways depending on where one comes from. For example, although the sight of a black cat is considered a lucky event in Britain, it is considered unlucky in many other countries. Dragons are viewed positively in China, but negatively in Europe and North America.

These examples illustrate dissimilarities between cultures that are both large and simple in the sense that they focus on a single cultural aspect that keeps the same meaning regardless of context. As a result, such variations in communication will often be identified on the spot. By contrast, subtle or complex differences are often identified much later in the communication process, when corrective action requires considerable effort and money. Sometimes, this realization takes place so late that there is not enough time to address it, resulting in a missed deadline.

In extreme cases, miscommunication can lead to casualties. For example, in the late nineties, a plane crash in the northeastern United States was caused, at least in part, by miscommunication between the pilot and air traffic controller. The plane was running short on fuel. But somehow the pilot did not manage to communicate the urgency of the situation to the air traffic controller, who put the plane on a holding pattern because of airport congestion. The plane then crashed when it ran out of fuel.

The ideas above are a small example of the importance of cross-cultural awareness for project success. (And) most of the chapters in the readings have touched on one aspect or another related to cross-cultural differences. Why? Because cultural diversity is all around us. We can no longer avoid the cross-cultural interaction that was once only the responsibility of *global* IT leaders. I cannot stress this enough, take time to research and acquire the cultural knowledge of your peers and/or business associates. It will make all the difference in the world when it comes to your project success.

⚠

IT Management Lessons

- Most projects fail due to lack of sufficient planning. Thoroughly identify and document your project goals prior to executing. This will help avoid any unpleasant surprises during project rollout.
- Make sure your staff is well aware if his/her exact role in the project. Communicate often and provide a detailed status report as the project progresses.
- Stick to all documented goals. Avoid straying off course from all planned and communicated tasks. If by some unfortunate

reason the project goes over budget, make sure you get the proper approval before moving forward.

- Cultural diversity can play a major role in the success of a project. Make sure you are culturally aware of the environment when projects stretch over neighboring borders or abroad.

Part X: Personal Management

> *"If you choose not to decide, you still have made a choice."*
>
> *-- Neil Peart*

Time Management

Throughout my career, one of the biggest challenges I see people struggle with is finding time to tackle the many responsibilities that go with an upper management position. No matter how masterful one may be at multi-tasking or delegating, it seems there is just not enough time in a day to accomplish all that is planned. Staff issues, business meetings, travel schedules, and any other job requirement you can think of will not go away. The only solution is to keep brushing up on your time management skills.

From a work perspective, there are many priorities that dominate a manager's day. Many are daily routine tasks that require a great deal of attention. Others are unforeseen interruptions that can totally disrupt what seemed to be a well-scheduled day. However, there are many ways of addressing each and every time disrupting issue.

Delegating is an important tactic, not only to help you with your workload, but also for developing staff, and creating an atmosphere of employee empowerment. The critical component of delegating is to make sure each employee knows his/her degree of autonomy and authority. If parties are not clear, MORE time is wasted through delegation. Below are some delegating options:

- Subordinate has full authority to make a decision without consulting boss.
- Subordinate makes decisions, but informs boss and anyone else involved, preventing any surprises or unexpected problems.
- Subordinate recommends a final decision, which boss must approve.
- Subordinate presents alternative solutions to boss, who makes the decision.
- Subordinate presents relevant information from which boss narrows down feasible alternatives. Boss then makes final decision after consulting with subordinate.

Each environment and situation will warrant its own option. Think carefully about your time objectives and analyze the availability of your staff. Furthermore, to avoid task related issues, be sure to identify the importance of the task before applying any of these options. Too little or too much delegation could result in unfavorable task outcomes, especially if it is related to a high profile project.

Using Talent

Organizational talent is the sum of skills and abilities available in your work unit. Effective use of organizational talent can save time and reduce frustration, while misuse results in the opposite. Here are some tips:

- Identify your own strengths and weaknesses. Try to be as objective as possible. When you undertake tasks in areas where your skills are not highly developed, you spend MORE time than would a person with more developed skills. By recognizing your own lack of expertise, you can avoid spending time unnecessarily. If you are not good at something, consider finding someone who IS good at it.
- Certain tasks can not be done efficiently by only partially trained or educated
  personnel. For example, web based data design work can be done internally, but only at a significant cost in terms of time and quality. While technology may allow us to do things otherwise reserved for specialists, that does not mean we can do it well, or do it efficiently. Unless you have other reasons for doing so, do not take on "specialist" type tasks when people outside your organization may be able to do the task more efficiently. For example, fiber optic infrastructures and routers require proper, specialized installation and configuration in order to function in the most efficient manner.
- Working with outsourced individuals can be time-consuming. Look to create a relationship with specialists so that they understand your needs. If you contract for service, do not always go for lowest price. Indicate that you are looking for a long-term relationship, and that you expect the specialist to save you time, not cost you time.
- Where you have a recurring need that you would like to handle internally, invest in your staff. Provide proper and ongoing training so that the person can become exceptionally good at it, rather than just mediocre. Only the "very good" will work efficiently, and the mediocre need constant support, which is time consuming.

- Return calls when it is unlikely that the other party will want an extended
conversation. Before lunch and towards the end of the workday may be good times. When calling, say: "I know you must be heading off to lunch, but I wanted to make sure I talked with you about..."
- Schedule meetings with a distinct termination time. When scheduling indicate this termination time to the other person, and/or ask how long the person needs. Stick to the termination time, and people will catch-on that you are serious about it, and will modify their behavior to fit the time constraint.

Managing Interruptions

Interruptions, be they on the phone or in person can be frustrating and time
consuming. Apart from the time spent ON the interruption, it may take time after the interruption for you to regain your original level of concentration and focus. The following suggestions should help make your life easier:

- When scheduling meetings (i.e. in your office), schedule them in blocks. Do not have one here and one there, but consolidate them, one after the other if possible. This will help keep each individual meeting to a reasonable and pre-defined length. Inform secretary or relevant people when each meeting will end and make it clear that you do not wish to be interrupted, and when you will be available.
- If you are constantly bombarded by random phone calls and visits, set aside a time each day (quiet time, focus time) to work on specific projects. Make sure staff are aware that this time is private and should not be intruded upon unless there is a dire emergency. Consider scheduling this time at the same time each day.
- If you have a "screener" who deals with visitors and phone calls before they are handed to you, make sure that they know what people should be screened to you and which people will receive return calls/visits. You do not NEED to see or talk to people, especially vendors, every time THEY want you. You can exert some control over the process.

- Set aside particular times each day to return calls. If you have a secretary inform him/her when you will be returning calls so this information can be passed on to the caller.

Taking Control

Work related tasks require a great deal of your time, but add to the mix some personal life issues and it creates even more unnecessary overhead that can overwhelm the most organized of people. The following are some of the most helpful suggestions for gaining control:

- *Take care of yourself.* Healthy eating habits, following a regular sleep routine and exercise all contribute to your well being which allows you to perform at your best.
- *Do the worst first*. If you find you cannot stop worrying about a certain task, then do the worst task first. Once that task is completed you will feel relieved and able to concentrate on the rest of your tasks, one at a time.
- *Focus on the task at hand.* Attempting too much at once and underestimating the time it takes to do it will surely put you right back in the center of feeling overwhelmed and a step behind.
- *Control the phone*. Decide when and where you will answer the phone. During off-time, use voice mail.
- *Separate work from play*. Are you running the family conversation the same as a meeting of the IT leaders? Think about it.
- *Keep a Master List*. Merge all your to-do lists, schedules and activity lists into one Master List. Use the technology that works for you whether it is index cards, spiral notebook, computer or a PDA.
- *Clean off your desk*! Clutter is a distraction and time waster. How much time do you waste looking for documents and files? Stay organized!
- *Learn to say no*. Take a reality break and identify what you can reasonably expect to get done. Accept your limitations and control your expectations. Would you really expect anybody else to work as hard as you do?
- *Use technology to help you*. Create systems for handling your routine tasks. Take the time to use the technology you manage to improve your own efficiency for performing routine business tasks. Use/create web pages, templates, or

databases for managing inventories, files, and e-mail communication.

- *Slow down.* Are you rushing through everything? Stop, look and listen.

Making the commitment to take control of your time is the first step to successful time management. What is the next step? Planning and prioritizing your time by putting your life goals in writing. Just thinking about what needs to be done does not cut it in our busy lives where everything is competing for our attention. Not only is having a written time management plan more effective because it requires you to concentrate on the tasks needed to carry out your plan, but putting it in writing also deepens your commitment to your goals. In addition, you will have a visible record to guide, track, and analyze your goals. Ideally you should have only one time planning system which bundles the following elements together for easy reference and portability.

Λ

> *"When we change our perception we gain control. The stress becomes a challenge, not a threat. When we commit to action, to actually doing rather than feeling trapped by events, the stress in our life becomes manageable."*
>
> *-- Bob Nelson*

Stress

Stress is all around us. How we deal with it is the question. Most of our stress usually comes from work related issues that stretch across our personal live. Technology itself provides some major challenges, however, add personnel and business issues and you have a recipe for a strong stress cocktail. And as an IT leader it will often be difficult to leave the baggage at work. The key is learning to manage the stress before it becomes overwhelming. The following are some suggestions to overcome Stress:

- Plan ahead. Ask questions about responsibilities, deadlines, and expectations. Be positive. Questions should show that you want to do the job better. Speak up is the workload is overwhelming.
- Avoid office gossip and constantly negative people. Too much negativity affects morale and is contagious.
- Seek out people who can provide positive feedback or encouragement.
- Set Priorities. If e-mail is essential, set aside time for it but answer in a few short sentences and try to check it only at select points during the day.
- Refocus. Look at what you contribute to be working, whether it is a larger goal or your family's well-being.
- Do not lose your confidence. Technology leaders sometimes under-rate their knowledge and skills. Sometimes it is only when they lose their job that they discover exactly how much they really know.
- Break up your work into smaller chunks. Complete a job before you start the next one. The satisfaction of seeing a job done will help reduce stress.
- Be clear that your job is not you. You are a person with a job. Remember everything else in your life: friends, hobbies, family.
- Reward your accomplishments. It is amazing what a little praise will do.

- Get enough sleep. Trading a half-hour of TV for extra sleep can change attitudes. Avoid late night computer activity.
- Set aside time to play. Play is an important aspect of being human, even if it is as simple as watching a movie with your children.
- Pay attention to your social life. Talking to people who care about you is an important safety valve.
- Change your scenery. Go to another room or take a walk outside. Take time off. Schedule vacations or even just a mental-health day before you get overwhelmed or burnt out.
- Take physical exercise, even if it's only walking. Your physical health affects how you experience stress.
- Lastly, eat right and drink plenty of water. This is an area most of us take for granted and forget to focus on. Junk food can add to an already irritating day.

There are obviously many ways to manage stress. To summarize, it is those things we cannot control that tend to cause us the greatest stress. By creating options for how to get things done, you gain a sense of control over your circumstances. Focus on what you can do, not on what you cannot do. You can only do one thing at a time, so identify the most important thing you need to do and then work on it. If your highest priority work is a large task, break it down into smaller more doable tasks. If you get bogged down, take a break from your most important work to focus on Priority 2.

Most irritability and health problems are usually related to some type of stressful issue. Gain perspective on your job responsibilities and take a break from time to time. This will renew your energy and allow you to stay keep focused on the stressful task at hand. Note, stress can have alarming symptoms. Some of these symptoms can also be caused by physical conditions. If in doubt, consult your doctor.

Team Stress Technique

When stress affects one of your team members it can have a negative affect on the rest of the team. The work place becomes counter-productive as teammates avoid or challenge the person. It can even affect the customer service image of your department if the person exudes the stress throughout the user community.

I am not sure where I first heard about this technique, but I use a strategy that helps diffuse a stressful work environment when someone is having a bad or stressful day. We basically created what I call a

"Bad Day Board". The board lists everyone's names with a magnet that could be moved to indicate the person is having a "bad day" from stress, personal problems, difficult users, etc. Initially meant to serve as a warning system for others, the group took on the challenge of trying to cheer up anyone who was having a "bad day". The strategy has become an important tool in maintaining a productive environment and the team states they have a lot of fun in the process!

⚠

"Good humor is a tonic for mind and body. It is the best antidote for anxiety and depression. It is a business asset. It attracts and keeps friends. It lightens human burdens. It is the direct route to serenity and contentment."

--Grenville Kleiser

Maintaining your Health

The majority of this book contains concepts and ideas aimed at helping IT people become better managers and leaders. However, all the knowledge in the world cannot help you become a more efficient leader if your physical health is not in line with mental demand of this type of position. When we feel irritable and drained our performance literally suffers. There is nothing like a well conditioned, well rested mind to handle the many complex issues and projects that await an IT manager.

As mentioned in the preface of this book, many people ask me how I maintain my energy levels and positive appearance, especially after long troubleshooting nights and long stressful business meetings. For the past twenty years my answer has always been the same, "Believe me, it's not easy! I basically make time to workout". If I go more than two or three days without some type of physical exercise, especially weight lifting, I tend to lose focus and my attitude degrades down to frustrated mode. Taking the stress out on the weights is the only way to real reset my mind. It prepares me for another challenging day of work and personal life.

Exercise

It is a well know fact that exercise is good medicine for stress relief and prevention of health problems. However, no matter how compelling the reasons to work out, sustaining a regular exercise program can be difficult. Excuses range from the plausible ("Exercise is too boring") to the absurd ("I'm in no shape to exercise"). People on a busy schedule cannot find the time, and those who have the time will not admit it.

In this day and age there is an abundance of statistical information that justifies the need to stay active. From staying physically fit for appearance purposes to flat out saving your life from diseases. Earlier in the chapter I mentioned the stress factor and listed many ways to manage it. However, in most cases, issues surrounding stress are relevant to the amount of mental pressure one can sustain. Regular exercise maintains and builds mental stamina, which helps fight stress. Here is a more detailed and basic listing of what exercise has to offer:

- *Strength and fitness.* Moderate exercise helps build strength, endurance, and flexibility of muscles, which will contribute to the health of muscles and bones. It also promotes cardiovascular (heart-related) fitness and endurance. Exercise may help promote flexibility and improved daily physical stamina. Vigorous activity improves fitness even more.
- *Weight control.* Obesity is a risk factor for heart disease and a culprit in other diseases as well. Exercise coupled with a healthy diet is the key to effective weight loss. For example, losing 1 pound of fat per week requires a deficit of 3,500 calories a week. Achieve this by eating less and exercising more. Drink plenty of water, at least six to eight 8-ounce glasses a day (more if you exercise regularly). Eat at least five servings of fruits and vegetables each day. Decrease fat intake to 30 grams or less per day. No more than 30% of your total calories should come from fat (10% from saturated fat). Include a lot of fiber (from fruits, vegetables, and whole grains) and complex carbohydrates (such as pasta, rice, or potatoes) in your diet.
- *Mental health.* Physical activity can improve self-esteem and impart a general sense of well-being. Exercise helps promote mental stamina.
- *Disease prevention (i.e. heart attacks and strokes).* The American Heart Association has identified physical inactivity as a risk factor for cardiovascular diseases (primarily heart attacks and strokes) which are the leading causes of death in the United States. Certain risk factors for cardiovascular disease can be reduced with exercise. These include high blood pressure, high cholesterol, obesity, and inactivity. Regular exercise can lower your risk of dying from cardiovascular diseases by 40% (1). Quitting smoking is also important.

Now the tough part, how do you stay motivated and keep from making excuses for exercising. First, you need to select an exercise program that is right for you. My routine generally consists primarily of weights, along with sporadic cardiovascular workouts. This, however, is not the best routine to follow. Factors such as age, weight, and physical conditioning all have an influence on the right type of workout routine you should follow. Second, the way to results is dedication. Here are proven strategies to help motivate you to start exercising and stay with it.

- *Develop an exercise habit.* It takes weeks to form a habit. So keep at it, knowing the more consistent you are in the beginning, the more fixed your new activity will become.

- *Maintain a "to do" list.* Reserve a time slot each day for working out, and do not let anything interfere. Not setting a time leaves you vulnerable to trying to find the time, which typically does not work. The best time to exercise is the most convenient time for you. Although you may be a "morning person," if mornings are too busy, you simply will not work.

- *Do not let others distract you.* Inform everyone of your exercise time and that you would appreciate their respecting your choice. When approached, invite others to either come along or come back later.

- *Be patient with yourself.* Some days you will be more motivated or have more time than other days. When possible, do more. When you do not feel strong, do less, or do something different. When you cannot exercise for a while because of illness, injury, or demands on your time, back off without guilt. A brief period of not exercising is not all bad. In fact, brief periods of rest can actually avoid workout burn out.

- *Stick to your plans.* Be prepared to exercise. It decreases the inertia of getting moving when demands arise. If you intend to exercise when you get home from work, for example, change immediately into your exercise clothing.

- *Team up.* Exercising with others can motivate you when you would rather not. But it can have a down side. A less motivated or less optimistic partner, for example, can drain you. An option is to have an exercise partner/spotter only once or twice a week (on weekends, for example), and to exercise alone the rest of the time. Choose the approach that works for you.

- *Set achievable goals.* The more easily you accomplish your goals, the more likely you are to sustain them. Set goals that emphasize the process (for example, exercising daily for 1 month) as well as the product (for example, jogging 3 miles in 30 minutes). When you achieve a goal, reward yourself. Decide on a reward ahead of time to motivate you on.

- *Do not forget to have fun.* Working out should not be all pain. When a body is conditioned, exercise actually feels great. Customize your approach to make exercise more enjoyable. For instance, read, watch TV, or listen to your favorite music while pedaling a stationary cycle. Or listen to some heavy metal music when lifting weights.

- *Affirm your goals.* Your subconscious believes what it hears, without reasoning. Affirm out loud each morning (when no one is listening!) that you are vibrant and looking forward to a chance to exercise. Then, when the opportunity for exercise arises, your mind will encourage you.

- *Listen to your body.* If you exercise regularly, your body may at times say no. Take the hint. You may be doing too much, and overtraining can dampen enthusiasm, causing you to quit. Shift to a milder form of exercise, or take a break. Know your body and do not work though pain. The old adage, "no pain, no gain" is not always a good approach. Believe me, 10 years later and my old shoulder injury is still plaguing me. Workout smart!

- *Complement exercise.* In addition to exercising, be sure to eat a low-fat, balanced diet, sleep well, and reduce unhealthy influences like smoking and alcohol.

Eating Right

Another well known fact is that in order to obtain the full benefits of exercise, the workout program must be accompanied by a well balanced diet. For the most part, I believe many people today are now trying (more than ever) to correct their bad eating habits. Much of the turn around is simply due to the many publicized studies that attribute serious diseases to certain types of food, as well as the quantity consumed.

Preaching about the problems with fast foods would be overly redundant in this day and age. Yet, today's busy lifestyles have made fast food the mainstay of all diets. Not to mention that it can also get pretty expensive. The only solution is to plan your diet carefully, even when traveling. How? Try the following:

- Choose a restaurant that serves a variety of foods and is willing to accommodate special requests, such as smaller portions.
- At your office, keep take-out menus of the restaurants that serve your lighter favorites so you can call in your order and pick it up on the way home.
- If you plan to splurge, eat lighter at your other meals that day or the next.
- Ask for salad dressing, sauces, and gravy on the side so you can choose the portions. Then dip fork tines in the sauce instead of pouring it over the food.
- Choose foods that are broiled, baked, grilled, roasted, steamed, poached, lightly sautéed or stir fried, or prepared with no more than a tablespoon of olive oil.
- Eat sparingly on a special occasion, or split with a friend any dishes that are fried, breaded, scalloped, au gratin, or alfredo.
- Try to eat the same portions you would at home. If the restaurant meal is larger, put the extra food in a doggie bag before you start eating, share it with your dinner partner, or ask for a split order. The same goes for fast food. Steer clear of "jumbo," "giant," "deluxe," and "double." Instead, go for a regular or junior-size sandwich.
- You do not have to order an entree. Appetizers often come in low-calorie options. Add a salad or soup, and you have a meal.
- If you must have dessert, order one with a fork for every member of your dinner party.

Creating a list of things to do (and not to do) is simple. The hard part is practicing what I preach. I do try to follow my own guidelines for eating right, but it is difficult to adhere to something when the temptation around me is so strong. I enjoy all the fatty and calorie ridden foods as much as the next person does. However, I have been able to remain successful in my diet and exercise routines because I fortunately have discipline for limiting the quantity I consume. This basically means that it is sometimes okay to cheat, as long as you do not over indulge.

With the proper approach, it is possible to maintain your nutritional regimen, even when you are away from home. By adhering to the simple rules above, you can stay the course and maintain a healthy weight, regardless of where you may be dining.

> *"Never mistake knowledge for wisdom. One helps you make a living; the other helps you make a life."*
>
> *- Carey*

Cross-cultural Fitness
Relaxation
Throughout my travels, one of the most interesting things I learned was the importance many cultures place on relaxation. For example, Asian cultures in San Francisco use their lunch hour to take part in group relaxation programs. Some of the most popular include:

- *Autogenic relaxation.* This is a form of passive relaxation where you do not move anything. You basically imagine different parts of your body becoming heavy, warm and relaxed.
- *Progressive relaxation.*   This form involves physical movement. You tense particular body parts, one at a time (your hand, your forearm, etc), hold the tension and then let it go (either quickly or slowly). You attend to the feelings of tension and relaxation so you can become better at identifying your tension, and learning how to release it.
- *Guided imagery relaxation.* This form involves imagining yourself in soothing, calm surroundings (on the beach, for example), and attending to the details of the experience (e.g. the sand on your feet and the warmth of the sun).
- *Meditation techniques.* Meditation for relaxation involves focusing your attention on a word, your breathing, or a simple object. Its purpose is to calm the mind.

A combination of exercise and relaxation can help us effectively face the pressure that is on all fronts; job, financial, political and many other sources. It is unfortunate, however, that so many people rather depend on anti-stress medications than to put forth a little effort on exercise and a little time on relaxation. The use of medications can inevitably result in other dependencies, which in the end will only increase stress levels and affect health.

Step it up a Notch
There are many ways of fighting stress and its effects, ways that will serve to improve your health and quality of life, rather than place it at risk. One of the concepts that most all cultures maintain, especially Chinese, is that inactivity is the major cause of illness. Thus, they have

developed a system called Tai-Chi that provides the benefits of working out without the use of weights. It brings into play every part of the body and benefits all bodily parts, not just the musculoskeletal system.

Tai-Chi has many distinct advantages over other types of exercise. Most systems of physical fitness only service part of the body. They concentrate on certain muscles or muscle groups, while neglecting others entirely. In addition, studies have shown that tai chi stimulates the central nervous system, lowers blood pressure, relieves stress and gently tones muscles without strain. It also enhances digestion, elimination of wastes and the circulation of blood. Tai-Chi uses every part of the body and incorporates slow, fluid movements with rhythmic breathing patterns. Some of the motions are swift, but these are short and require very little effort. You can hold a position for some time, but not for too long. One is usually in a state of meditation while doing the exercise.

Learning Tai-Chi is not a quick process. It is about patience and repetition. If you are just starting out, then patience will be your biggest requirement. Start with learning a move, practice it until you get the hand of it, then add a move a few days later. Practice the two moves together, then learn the next move, and so on, gradually building up. The more the repetition, the more your body becomes in tune, and it (as well as your mind) remembers the movements.

Certainly, I am not an expert on Tai-Chi. If you are interested in learning this popular moving meditation technique, there are many books out there that provide simple descriptions of the movements. The following is a list of some important things to keep in mind when learning each movement:

- Always keep your head straight while you are doing Tai-Chi. This aids balance. The chi node at the top of your head is a central meeting point of nearly all the chi meridian lines in the body, and is where chi is released upwards into the atmosphere.
- Keep your back straight. You will form a vertical "plumb line" from the top of your head to your tan tien (your centre of gravity) and of your chi (about two inches below your navel).
- Splay your toes slightly. This really does make a difference to the way that you balance, especially at the times when you are on one leg.
- Always practice in fluidic, constant motion. You do not do one move, then another, as separate things, "stopping" between each. The end of one move should flow seamlessly

into another. While you are finishing one, a part of your body should be beginning the next.

- Make hand movements circular in motion. This increases the fluidity of the motion, and brings expression into the form. Breathe out as you push out, and breathe in as you bring your movements in.
- Always be relaxed, especially in your arms. Be wary of pains, especially in your knees. If you feel pain ease off a bit.

The fundamental tip for doing Tai-Chi is to do it often. Practicing Tai-Chi only once a week will provide you absolutely no benefit. Make a practice of doing it every day, or at least four to five times a week. Set a little quality time aside each day, and know that this is the time that you will be doing it.

Λ

> *Wisdom is not a product of schooling but of the life-long attempt to acquire it."*
>
> *-- Albert Einstein*

My Personal Workout

It is difficult to detail exactly what and when I workout because my routines are usually based on my mood and physical ability of that day. The point to remember is it does not have to be a complicated or overly difficult workout. As long as you put some effort into it, the basic exercises will help you maintain and/or improve your physical appearance and overall health. Remember! Patience and dedication are the cornerstones to meeting your overall health objectives.

I basically warm up with a five minute abdominal workout, which consists of the following example exercises:

1) *Seated Knee-Up (lower-ab region).* Sit on the edge of a stable chair or bench, grasping it at your sides to stabilize your body. Lean back slightly and extend your legs down and out in front of you. With your legs together and toes pointed, bring your knees in to your chest in a smooth motion as you curl your upper body slightly forward at the same time.

2) *Twisting Crunch (upper-ab & obliques).* From the lying crunch position, hands loosely placed behind your ears, cross your right elbow toward the outside of your left knee in a smooth motion. You do not need to actually touch your elbow to your knee. Alternate sides after each rep.

3) *Crunch (upper-ab region).* Lie on the floor with your knees bent at about 60 degrees, feet flat on the floor about shoulder-width apart. With your hands lightly gripped behind your head (not pulling), curl forward to bring your shoulder blades just off the floor.

Do as many reps of each exercise as possible, starting with the first exercise, then doing numbers 2 & 3 respectively. Arrange exercises as a circuit resting 10 seconds between sets. Do the circuit a total of three times. During your next workout, do three straight sets of each exercise (not a circuit), resting 25 seconds between sets. The point is to avoid redundancy.

There are many variations of abdominal (and other muscle) workouts. Before I list the rest of my routine I want to remind you of some basic ideas.

- Research and find different exercises that target each body part. Sort them out and create routines that work for your schedule. There are many videos and books out there that provide routines that are either too intense or require a great deal of workout time. Creating your own routine will make your workout more enjoyable.
- If you are working with weights, keep your routine down to under thirty minutes. Unless you are a body builder, twelve sets per muscle is over kill. If you want a longer workout, add cardiovascular exercises to your routine (i.e. aerobics, bicycling, and jogging), for another twenty or so minutes.
- Change the order, type and routine of your exercises on a regularly basis. The body can quickly adapt to repetitive routines, which hinders muscles growth.

After my warm-up, the remainder of my daily workout will consist of weight training. To avoid complicating the description of my daily or weekly routines, I am only providing a general overview of the exercises I use the most, including alternate exercises for those who do not have that specific exercise machine at home. Later in the chapter I will provide some different variations or techniques which can help shock the muscles without changing the exercises. In the meantime here are some basic movements:

Back

- *Pull-Down to Front.* Grasp the bar with a wide overhand grip. From an arms-fully-extended position, pull the bar toward the top of your chest, elbows coming down and back. Squeeze your shoulder blades together before slowly returning to the starting position. Avoid not keeping your lower back slightly arched throughout. Alternate exercise: One-arm dumbbell row (or pull-up, if you have a bar).
- *Back Extension.* Lie facedown on a back-extension bench with your heels under the pad. With your body in a straight line, head neutral, lower your torso to about a 90-degree angle to your legs. Raise your torso back up to the starting position in a smooth motion.   Alternate exercise: Lie on the floor and, with

your hands behind your head, lift your torso up a few inches, holding for a two count before lowering and repeat.

- *Bent-Over Barbell Row.* Keep a slight bend in your knees and your torso at a 30-45-degree angle to the floor, while maintaining the natural arch in your lower back by holding your head up. Grasp the bar with a shoulder-width, overhand grip and lift the bar into your lower midsection, moving your shoulder blades back on the ascent. Then reverse the movement, controlling the bar on the descent.

Biceps

- *Alternate Dumbbell Curl.* With your upper arms by your sides, palms facing in, flex your biceps on one arm to bring the dumbbell toward your shoulder, twisting your wrist as you lift so that at the top, your palm faces up. Reverse the movement, twisting your wrist back to the starting position at the bottom. Then complete a rep with the other arm. Completing the move for both arms constitutes one full repetition.
- *EZ-Bar Curl.* Grasp an EZ-bar with an underhand, shoulder-width grip, lock your upper arms by your sides and, without swinging from your lower back, curl the bar toward your shoulders. Squeeze your biceps before lowering the weight and repeat. Avoid letting your elbows flare from your sides for leverage.

Calves

- *Seated Calf Raise.* With your knees under the pads, place the balls of your feet on the edge of the foot platform. Release the safety lock and lower your heels until you feel a stretch in your calves. Push through the balls of your feet to raise your heels as high as you can. Squeeze your calves hard at the top. Avoid turning your ankles in or out; keep your toes and feet in their natural, relatively straight position. Alternate exercise: Place plates on your lap to provide resistance, or perform additional sets of standing dumbbell calf raises.
- *Standing Dumbbell Calf Raise.* Stand holding a dumbbell at your side. With the ball of your foot on a raised surface such as a step and your leg straight, lower your heel to stretch your calf, then press up on the ball of your foot as high as you can. Complete all reps for one leg before switching legs.

Chest

- *Bench Press.* Lie face up on a flat bench, feet planted on the floor, with your arms perpendicular to your shoulders. Bring the bar down and out slightly to mid-chest level, pause, and press back up, straightening, but not locking, your elbows at the top. Avoid raising or using your pelvic area when pressing.

- *Flat-Bench Dumbbell Flye.* Lie face up on a flat bench, holding a pair of dumbbells at arms' length over your chest. Your palms can either face each other or you can do the flye with palms-forward, which may take some stress off your shoulders. With your elbows slightly bent and pointing out, and locked in that position throughout, lower the dumbbells in an arc out to your sides until you feel a good stretch in your pecs. Reverse the motion, but do not let the dumbbells touch at the top before beginning the next rep. Avoid not keeping your elbows out to the sides to recruit the chest. Bringing the elbows closer in to the sides gets your triceps more involved in the movement than you want them to.

Shoulders (delts)

- *Military press.* Sit down at the bench press and plant your feet, thighs parallel to the floor. Reach up, lift the bar from the supports and lower it straight down and directly in front of your face, touching your clavicles. Your elbows should point straight down. Pause briefly before reversing the motion, stopping just short of elbow lockout at the top.

- *Dumbbell Lateral Raise.* Stand and hold a pair of dumbbells in front of your thighs, palms facing each other. With your elbows slightly bent and locked in this position, lift the weights up and out to your sides until your hands and elbows are about shoulder height. Avoid leaning backward as you lift to generate momentum

Legs

- *Leg Extension.* Adjust the seat back and footpad so your knees are flush with the edge of the seat and your ankles are just below the footpad. Grasp the handles to keep your body stationary and press your shins against the footpad until your legs are almost fully extended. Squeeze your quads briefly, then slowly return to the starting position and repeat. Avoid snapping up into the top position, putting your knees at risk of hyperextension. Alternate exercise: Dumbbell lunge.

- *Lying Leg Curl.* Position the footpad so it presses just above your Achilles tendons, and lie facedown on the machine. Grasp the handles and bend your knees to curl the weight toward your glutes. Squeeze your hamstrings at the top, then slowly lower the weight and repeat. A good mental trick to activate your hams is to imagine them flexing on the ascent; as a visual, think of your hamstrings like your biceps flexing on a curling exercise. Avoid not taking the movement through a full range of motion. Alternate exercise: Dumbbell lunge.
- *Barbell Squat.* Step under the bar in a shoulder-width stance, and keep your elbows back to form a ridge along your upper back where the bar can sit. Take a deep breath and, keeping your head up and entire body tensed, especially your abs, bend at the knees and let your glutes track backward to lower yourself. At the point where your thighs are parallel to the floor, reverse direction, driving up forcefully through your heels to a standing position. Avoid using plates under your heels, and rounding your lower back as you move through the range of motion.

Triceps

- *Dumbbell Overhead Extension.* Sit on a flat bench or a low-back chair. Lift a dumbbell overhead, cupping the inside edge of the top of the weight with both hands with a tight grip. From an elbows-extended position, slowly lower the dumbbell down behind your head, stopping just short of allowing the weight to touch your neck. Then reverse, squeezing your triceps when extending your elbows fully at the top. Avoid letting your elbows flare out to your sides too much; try to keep them facing forward.
- *Lying EZ-Bar French Press.* Lie face up on a flat bench and grasp an EZ-bar slightly inside shoulder width. Keeping your upper arms almost perpendicular to your torso (angled just slightly back), lower the bar toward the top of your head by bending your elbows. Press the weight back up in a controlled motion and squeeze your triceps at the top. Avoid not keeping your upper arms locked in position; your elbows should serve as a hinge.

Variations

When possible, I apply a three day on one day off approach to working out. However, most of the time I only manage to hit the weights two consecutive days without an interruption. My first goal is to get to every muscle at least once a week. The second is to change the type of exercise and routine at every workout. Some examples of my routine variations include:

- Day one I will focus on all pushing exercises (i.e. chest, shoulders, and triceps), day two I will focus on the pulling exercises (i.e. back and arms), and day three I will focus on legs (only).
- For those who really want to see muscle growth results. Try a split routine, where you work only one body part in the morning and another body part in the evening. This allows you to rest and blast each muscle with the same intensity.
- For the very busy, try to get to the main muscles at least once per week. Bench press, back rows and squats will help maintain your physical shape without taking too much time from your schedule.

In addition to routine changes, I often modify the number of sets, reps and amount of weight for each specific exercise. The following are some examples of these variations.

- The standard for gaining and maintaining muscle mass is to use heavy resistance. But not too heavy. As the old adage goes, "slow and steady wins the race." It takes time, patience and dedication to gain muscle. The basic method for steady increases is to use a weight that represents 70 percent of your overall maximum lifting weight for any given exercise. That is, a weight that allows you to do anywhere from five and ten reps. I do a total of three to four sets for each exercise using this routine.
- Then there is the cardiovascular weight workout. This involves using a weight that represents approximately 50 percent of you maximum lifting weight for any given exercise. A weight that allows you to do between seven and ten reps. The key to this routine is rest no more that 15 seconds between sets. I do an average of four to five sets per exercise.
- Big weights mean big muscles and the best way to get there is to use the powerlifting method. A weight that represents 80 to 90 percent of your maximum lifting weight for any

given exercise. Here I apply less reps (two to four) and less sets (two or three) per exercise. My rest intervals are also much longer (at least one minute).

These are the basic exercises and guidelines of my workouts. Whether I am on the road or at home, I focus on making every workout count and I let my body tell me what muscle I should focus on. That is, if my legs feel thin I focus on legs, if my shoulder hurts I avoid exercises that require delts, and if I feel strong I will focus on a heavy bench press workout. As a finale, I am add a small tai chi routine to the end of each workout. It really helps me calm down after an intense weight session.

Take some time and find a good routine that works with your schedule. If weights are not your cup of tea, find some type of cardiovascular exercise that involves resistance. After you have identified your routine, write down a few different variations that you can apply at home, on the road, or even in the office. The last important thing is to commit to your goals today. I hear everyone say, "I'll start Monday". That type of mentality is only a prolonged vision. In the IT field we never know what Monday will bring. Therefore, take advantage of any spare moment you may have and start your routine.

The bottom line for sharing this information is that I depend on my workout time to burn not only calories, but also the frustration and stress of work and everyday life. It helps me stay sharp and focused. IT management requires a great deal of personal investment that rarely comes with tangible compensation. The ones who suffer are the loving family members who rarely see you and yourself, who in the long-run can physically and mentally deteriorate. The only way to maintain a good mental balance between work, play and family is to exercise regularly. Give it a try! You will find that many if your current, overwhelming priorities will seem more feasible and manageable.

⚠

IT Management Lessons

- Effective time management is crucial to your IT management success. When you feel overloaded, use your delegation skills to relieve some of the workload.

- Stress can cause your whole management sphere to come crashing down. Do not let the stressful requirements of your duties take over your world. Take control, stay organized and relax!
- Many countries have developed great relaxation techniques to relieve stress. Research and try different routines that help relax the mind.
- Create an exercise routine that works for you. One that helps you adhere to your busy schedule, yet still gives you the resistance workout that your body needs. Gyms are great. However, often times they are over crowded and distractive. If you have the means, invest in your own home gym or weight set.

> "I get a lot of credit for a lot of things that our people do... I wish the spotlight could be turned on them, individually and collectively, because they're the ones that do it for our customers. They do it every day...."
>
> --Sam M. Walton Founder, Wal-Mart

Summary

Change is inevitable! Technology will change, business requirements will change, culture will change, and ultimately you will change. However, the concepts that have helped managers maintain a successful IT/Business environment have continued to survive the test of time. Nevertheless, those management concepts are continually complicated by the introduction of technical applications and devices. Therefore, my advice to you is to "keep it simple". Avoid applying your technical engineering philosophies to your IT manager liaison functions.

Unless the company you work for is a technology company, technology should not be driving the business. On the contrary, the business drives most technology requirements. And as technology advances, an IT manager's responsibility of helping the business find ways to be more productive is becoming more important than ever. Therefore, keep an open mind about the information provided in this book. It should be used only as a foundation of your management approach. Modify, adjust and apply the concepts in a manner that will conform to your personality and domain standards.

Although IT and the business need to function as one smooth machine, they are actually two separate worlds when it comes to the employees; each requiring a different type of management and communication approach. That is, employees from the business side respond more effectively to business related development techniques. (And) technical staffs often require a more analytical and structured approach. Add the cross-cultural variable to your IT management mix and you have a recipe for one challenging position.

As an IT leader, it is your responsibility to make any separation between the business and technology seem as transparent as possible. Unfortunately, from a business management perspective, technology people tend to make many technical issues more complex than is obvious. The techy jargon and the frequent use of acronyms is highly effective only when technology people are speaking to other technology people. The problem is that many IT leaders find it difficult to gauge their audience and rationalize the use of technical elaboration when communicating with the business. Consequently, the separation between the business and technical staff often becomes wider. Therefore, pay attention to your communication style and avoid getting too technical during company conferences, meetings, staff performance reviews, or even day-to-day conversations with Sr. business management.

In closing I would like to reiterate some high points to help utilize the concepts more effectively:

- *Stay organized.* This will help simplify your life. You will get more done in less time.
- *Get to know your team.* Learn all you can about their culture.
- *Stay up to date on technical advances.* Keep up to date on current and upcoming events in the technology world.
- *Keep learning.* Maintain your management skills by attending as many training courses per year as possible.
- *Stay fit.* A tired mind and body can work against you.

To this end, I would like to say that it takes a special kind of person to manage a technology environment. It requires a continual progression of learning, both technically and administratively, in order to remain successful. Make your vision a reality by writing down your goals and executing them in a strategic and organized manner. Your attention to detail will be rewarded in many ways - a dedicated successful team, a stable systems technology environment, and a highly respected reputation for understanding and managing resources that are critical to the business. Thank you and good luck!

Recommended Readings

Johnson, Spencer. *Who Moved My Cheese.* G.P. Putnum's Sons, 1998

Blanchard, Kenneth. *The One Minute Manager.* New York: Berkley Books

Nelson, Bob. *1001 Ways to Reward Employees.* New York: Workman Publishing

Bossidy, Larry, and Ram Charan. *Execution: The Discipline of Getting Things Done.* Crown Business, 2002.

Brin, David. *The Transparent Society: Will Technology Force Us to Choose Between Privacy and Freedom?* Addison-Wesley, 1998.

Shapiro, Carl, and Hal R. Varian. *Information Rules: A Strategic Guide to the Network Economy*, Harvard Business School Press, 1999.

Chaney, Lillian H; Martin, Jeanette S.; Tung, Rosalie. *Intercultural Business Communication (2nd Edition).* Prentice Hall, 1999)

Lewis, Richard D. *When Cultures Collide*; Nicholas Brealey, 2000)

Shwalbe, Kathy. *Information Technology Project Management.* Course Technology, 1999

Brown, Stanley A. *Customer Relationship Management: A Strategic Imperative in the World of E-Business*. Price Waterhouse Cooper, 2000

> *"If it isn't a profit center, or something you want to be great at. Than it should be outsourced to someone who does want to be great at it..."*
>
> *-- Jack Welch, former Chairman GE*

Case 1

Comtech Applies IT Outsourcing Strategies to Cross-border Firms

By A. Mata, Owner and President of Comtech Technical Support Solutions, Inc.

Abstract

Information Technology (IT) is now a significant cost to most businesses, often enhancing their productivity and profitability. In fact, an average company currently spends 5 to 10% of their annual gross sales revenue on IT alone. Many of the higher expense percentages are associated to communication and technical support requirements for company's that have international and/or remote sites.

Today's borderland regions contain a large number of these cross-border organizations, better known as "Maquiladoras". (And) 80 % of these organizations depend heavily on technology for their business functions (i.e. production reporting, planning, and varied communications). Comtech provides IT solutions and support services from the United States to Fortune 500 companies; with extensive expertise and experience in providing high quality network infrastructures and TAC (Technical Assistance Center) support to manufacturing clients.

Through Comtech, customers benefit from increased purchasing power, access to a broad range of skills and expertise, voice technology support, third-party cost savings, and IT management that supports the overall business objectives. Tasks that are critical, but non-core competency - the jobs that must be done, but are not easy for employees or owners to learn - are the tasks that are most often outsourced. For example, desktop and server support, LAN/WAN maintenance, PBX switch administration, AS/400 device installations and programming, and website development are all things that may be done better and more efficiently by an outsourced specialist.

Successful delivery of IT systems is imperative for cross-border manufacturing companies. Unfortunately, many of the technology infrastructures located over the border are severely outdated or poorly managed; creating an unstable environment that may lead to production impacting outages. This downtime can be extremely costly for any business, especially one that is dependent on real-time online processing throughout the interconnecting sites.

Cross-border companies traditionally hire indigenous technical staff to administer their data center and user environment. Although there are many skilled technicians located in the international regions, this approach presents many risks to the operational functions of the domestic site. One of the most serious is the lack of standardized equipment and support methodologies, which can potentially compromise operations. Consequently, creating a wide variety of negative financial and business performance issues.

From a business perspective, the cross-border plant is considered a sister site for the domestic company and is treated as a full extension of the organization. However, from a physical operations standpoint the international sites maintain a much different technology support approach, which is often inconsistent with the host company's standards. Many of these differences are based on cultural work ethics and teaching principles.

This case study outlines Comtech's approach to enhancing international end customer service delivery through technology outsourcing. Customer support represents one of the most critical facets of customer relationship management for any product manufacturer. For high technology support companies, this constitutes a very significant portion of the overall service cost. Additionally, the statutory obligations of providing for end-of-life products compound the cost structure. This also demands considerable management time and effort.

Just five years ago international organizations were skeptical about outsourcing their technology environments to cost effective support companies. However, recent research indicates that 40% of Fortune 500 companies are now outsourcing their technology activities. This enforces the logical concept that any organization looking at reducing costs should entrust their technology support activities to a suitable outsourcing partner. This requirement is taking on greater urgency in the current times.

Also outlined in this case are the typical problems affecting cross-border customer support and a few solutions that can help control the situation(s). Presented are some operational details that any company seriously considering outsourcing their technology support department should keep in mind.   Furthermore, the readings addresses an approach to the outsourcing of a manufacturing company's technology environment, as well as providing effective customer support through a centralized help desk model. These items include:

- Defining the various levels of customer support and the problems unique to each level.
- Specifying the scope of Comtech's suggested assistance.
- Enumerating the advantages of using an outsourcing partner to address TAC support issues.
- Comtech's credentials for providing class leading TAC support.
- A brief sketch of the implementation plan. Listing a typical set of Service Level Agreements (SLAs) that could be employed in monitoring the performance of a support center.
- Other examples of Comtech's support experiences.

Need for Outsourcing

More and more companies are going over the border or offshore to manufacture and finish their products. (And) many if these firms have formed partner-ships with outsourcing vendors such as Comtech for their customer support requirements. In fact 40% of the Fortune 500 companies currently outsourcing have some type of cross-border operation.

The most important reason for outsourcing their technical staff is significant savings in time & money. Labor accounts for more than 75 percent of the engineering and support costs and finding professionals with the relevant skills is difficult. Recruiting, hiring, and training people to meet the constantly changing needs of the IT environment can be a costly investment. Companies on an average spend more than 10 percent of their budget recruiting and training new staff.

Over the past five years, companies have begun to outsource many of their internal IT services, such as help desks, data center support, software support and software development. The most popular cross-border location is El Paso, which combines high quality with low costs. IT companies located in the Southwest region of Texas now have excellent capabilities and are now in an ideal position to provide more value added services.

Although outsourcing to technology firms in other states may be possible, companies can gain a strategic advantage from the lower wages and benefits prevailing in that general region of Texas, where hourly, fully loaded rates can be 15 to 30 percent below those in other parts of the United States. Companies that use cross-border partners to do their non discretionary programming and networking (Maintenance, Call centers, Level-2 support) also find the lower costs free up capital for new development efforts of business strategic importance.

Benefits of Outsourcing

Economic turmoil has driven companies to find more lucrative ways of saving money in order to remain competitive or even remain in business. Short-term or quick savings projects are no longer justifiable methods of demonstrating financial stability. Both the company and the outsourcing partner benefit long-term from this type of arrangement. These benefits include:

- *Cost.* Cost savings of about 40% are easily realizable with a technology outsourcing partner. While there are multiple department options for outsourcing (i.e. HR, Facilities, etc.), technology is the only one that is high on salary expense savings and long-term cost effectiveness.
- *Lower Manpower costs.* Texas has one of the largest pools of technical skills available in the United States. Combined with a competitive wage structure in the region, the labor costs in El Paso are one of the lowest in Texas.
- *Lower Infrastructure costs per unit.* The outsourcing partner can utilize the resources very efficiently by leveraging on economies of scale. As the resource utilization increases, the partner is able to realize a lower infrastructure cost per unit.
- *Optimized ROI.* All of the customers' costs can be converted to variable costs which can be 40% lower than the current running costs. Moreover with little or no investments in infrastructure the returns would be significant. This is very essential in the current times when matching costs to real revenues is paramount. This also ensures that scaling the resources or ramping them down can be done at a predictable cost.

- *Management overhead.* One of the most important motives for outsourcing is to reduce the effort expended on cost centers. By outsourcing, this effort can be localized.
- *Management through SLAs.* The management effort is focused on the important things like the SLAs. The other mundane, but important, tasks are taken care of by the partner.
- *Infrastructure management.* While running a technology support operation, one of the most critical functions that need to function flawlessly is managing the infrastructure to ensure availability and planning for future requirements. This requires considerable amount of management time and effort. Outsourcing helps in focusing the organizations energies on core tasks and leaving operations management to the outsourcing partner.
- *People management.* Another area that consumes a lot of effort is managing the people that form the TAC. This task has multiple dimensions:
  - Training
  - Recruiting
  - Scaling up or down as per requirements
  - Matching the number of personnel available to the case load in any given shift

Challenges in Outsourcing

Outsourcing operations, be it support or development, brings with it, its own unique set of challenges. Listed below are some of the major issues that could be faced while outsourcing.

- *Partner's Technical Competence.* Technical competence is the cornerstone of any support operation. The partner's other strengths are worth nothing if there is no demonstrable technical competence to back them up. It would be ideal if the partner's technical competence is deep and backed by historical data enhancing end customer service delivery.
- *Reliable infrastructure.* It is crucial that the infrastructure, both communications and facilities, is reliable to ensure smooth functioning of a support operation like a TAC. The reliability of the infrastructure is influenced by the partner's competence in managing the infrastructure as much as it is by availability. In fact, a partner's ingenuity can overcome

gaps in availability. Expertise in setting up and managing reliable infrastructure is a key consideration in choosing the right partner.

- *Cultural Issues.* Cultural diversity is a problem that needs to be addressed by any TAC, as people from varied cultures would interact with a global TAC irrespective of its location. This problem is compounded slightly when the TAC is outsourced, as there is a perceived lack of control. A good partner should not only have sufficient experience in handling cross-cultural issues but also good internal training processes to effectively handle these issues.

- *Security.* Security is often a major stumbling block in outsourcing operations. This is often a mindset and trust issue rather than a competence issue. Any outsourcing partner would need to address security concerns to the complete satisfaction of the customer. Experience in providing security to information and facilities is a definite requirement for a partner.

- *Remoteness.* By definition, an outsourcing partner is remotely located. While sometimes this geographical distance can work as an advantage (for example, by enabling operations across multiple time zones), could pose a challenge in handling high severity cases that need proximity to the customer site.

- *Accessibility to skilled manpower.* Availability of skilled manpower is essential for smooth running of operations. Organizations spend huge amounts of time and money in hiring and recruiting and training the right people just to enhance their end customer service delivery. Outsourcing partners can provide a wide variety of technical people for any type of project.

- *Choosing the right partner.* While outsourcing has a lot of benefits, the partner chosen for outsourcing ultimately determines its success. With the right partner, the benefits can be maximized and the relationship symbiotic. Some of the aspects that would be needed to be considered in choosing the right partner are:
    - Experience: The partner should have the right breadth and depth of experience both technically and in management to deliver the commitments made.

• Financials: The financials of the partner are very important in determining their suitability. It is not just the raw numbers that are the determining factor, though they are very important. A degree of transparency is required in the financial management. A publicly listed company would be ideal as it is statutorily required to be transparent about its financials.

• Stability: A partner who has been in business for extended periods, with a long term outlook towards customer relationships is required. TAC support is essentially a long term commitment and the stability of the partner is a critical factor in determining suitability of a company as an outsourcing partner for TAC operations.

Choosing the right partner ensures maximum realization from outsourcing. Comtech has the right credentials to qualify as the best partner for a company looking to outsource Technology operations.

Implementation Overview

The following sections outline the generic overall *Operation Model* plan for transitioning the support for any service. Broadly, the steps are as follows:

- Identify the service for support
- Transition knowledge about the service to the partner
- Sign-off on SLAs & resources like people, technology equipment etc. necessary for supporting the service
- Parallel support operations for the service
- Fine tune operations to improve on the performance

Of these steps, the transition step is crucial to ensuring the success of this operation. These sub-steps include:

- The support personnel will be responsible for…
- Case receipt/Logging
- Case assignment
- Technical Troubleshooting
- Posting status updates on support tracking
- Provide technical tips, FAQs and checklists for efficient operations

- Simulation of the problem
- MIS reporting
- Escalations

Reporting

Given below is an indicative list of reports that can be produced. The actual reports that are needed should be finalized mutually. The following reports will be generated accordingly:

Shift end reports.
- Initial response reports
- SRs received by severity, priority & technician engineer assigned
- SRs closed by severity, priority & technician engineer assigned
- Unassigned SRs
- Equipment failure reports
- Time sheets for each technical engineer

Weekly Reports
- Analyses of SRs by engineer, severity & priority
- Analyses of SRs by geography
- List of rejected SRs & re-queued SRs by technical engineer

Monthly Reports
- Consolidated SLA reports
- Data center downtime
- Link Downtime
- Dropped calls report
- Invoices
- Status of defects raised by the TAC - resolved, closed, pending, overdue.

Account Review Reports
- Relationship dashboard

Roles and Responsibilities

The responsibilities of the key roles have been outlined below.

Customer support representative responsibilities
- Receive cases
- Professionally and courteously attend to callers
- Log all cases and provide callers with tracking numbers

• Provide as much help online as is practicable (With the intent of closing the case)
• Ensure complete satisfaction for the caller/customer
• Interface with customer for any help in simulation of problem
• Work on logged cases
• Troubleshoot the problem and inform the customer about the solution
• Log all information relating to a case (Caller, hardware details, software version etc)
• Check in the knowledge base for similar known problems and provide already decided workarounds and/or tips for resolving the issue.
• Post updates on call status
• Escalate unsolved problem to client's support personnel
• Track all calls to closure
• Track defects raised to the client's backend to closure

Technical Specialist responsibilities
• Expertise in a specific domain
• Acts as expert resource for support personnel to aid quick resolution

Staff Manager responsibilities
• Work allocation to and load balancing of support staff
• Load matching
• First level escalations
• Reporting
• SLA monitoring and control
• People management

Regional Manager responsibilities
• All support center operations
• Measure, monitor, report and take corrective actions on the metrics for any deviations from norms
• Monitoring support center performance
• High level reporting
• Planning for and procuring all needed resources
• Take part in reviews with the client management
• Ensure adherence to processes and practices
• Continuous improvement
• Second level escalations

Backup & Disaster Recovery

Backups should run after every shift. Backed up data should be moved to offsite locations daily. All critical equipment should have equipment level or card level redundancy. In the event of an equipment failure, services should be restored with no noticeable interruption (transparent to users). A communications link failure can be handled via the backup links on different media and providers with no loss of operations. Data failures (e.g. a virus attack) can be recovered from in about an hour. If there is a facility level failure, partial operation can be restored using other facilities, if so provided for.

Total Cost of Management (TCM)

TCM consists of both visible and invisible costs. Directly visible costs are (a) cost to resolve downtime (b) Technical support costs which includes the cost of help desk, support staff, tools, payments to vendors and (c) administration and management costs to supervise the support staff and to do vendor management. The other part of TCM which is usually not visible, but can be much more financial impacting, are the hidden costs. (figure 1.0). These hidden costs consist of (a) productivity loss of all affected employees due to the down time and (b) revenue loss to the organization because of the outage.

Figure 1.0

```
                              Cost to
                              Resolve
                              Downtime


              Technical
              Support Costs


              Administration
              and Management
              Costs


              Downtime
              Productivity
              Loss


              Downtime
              Revenue Loss
```

Conclusion

IT Outsourcing is gaining popularity as cross-border business managers, confronting lean economic times, take another hard look at what is and is not a "core competency." More than ever before, IT is under the microscope and financial auditors are reevaluating their TCM models. I am referring to the analytical processes that make IT seem insignificant within an organization. It is all done in an effort to ferret out "technology cost centers" so they can be pared back or winnowed out.

There are obvious gains in TCM that favor IT outsourcing standardized services. The economies come from the traditional focus that market specialization creates. A technology outsourcing company that specializes in IT infrastructure management can bring to the TCM table years of invaluable experience, industry certified expert skill, world class processes honed through practices, significantly enhanced service levels, at a cost savings that is passed on to the enterprise.

IT outsourcing makes sense when customer requirements exceed the provider's ability to meet and exceed those requirements. It reduces the hassle of providing a non-core service. It is also effective when an organization is looking to minimize risk while still achieving its growth goals. However, keep in mind that IT Outsourcing should not be considered a quick fix. It should be an integral part of a long-term strategy.

About Comtech

Comtech Technical Support Solutions, Incorporated is a rapidly growing international technology outsourcing company from Texas, providing innovative and cost-effective IT solutions and consulting. Comtech has 15 years experience in understanding organizations and designing, implementing and supporting systems to realize the true value of technology for any type of business; providing expert services in virtually every facet of information technology management, from implementing emerging technologies to supporting day-to-day IT operations.

Comtech provides international operations support services for all major hardware and infrastructure software platforms including mid-range systems, server platforms, databases, application servers, operating systems, IP-based networks, and wireless networks. Their technical staff provides full operational support including systems monitoring, network monitoring, problem determination, problem reporting, problem escalation, system upgrades, change control, version management, backup and recovery, capacity planning, performance tuning and system programming.

"Success in technology relates to the basic notion that one person, regardless of their education, does not make up a team and the best and most successful leaders had many talented people assist in their goals. Hence, the art of success through team synergy."

-- Albert Mata, MBA, PMP

The Foundation of Excellence

Tentative efforts lead to tentative outcomes. Therefore give yourself fully to your endeavors. Decide to construct your character through excellent actions and determine to pay the price of a worthy goal. The trials you encounter will introduce you to your strengths. Remain steadfast...and one day you will build something that endures: something worthy of your potential.

- Epictetus
Roman Teacher Philosopher
55-135 A.D.

References

Jackson, T. 1995. *Cross-cultural Management*. Butterworth-Heinemann: Oxford.

Robbins, S. (2001). *Organizational Behavior.* Upper Saddle River: Prentice Hall.

Krajewski, J. Lee, & Ritzman, P. Larry. (2000) *Operations Management: Strategy and Analysis* [UOP Custom Edition]. 5th ed. Boston, Massachusetts: Pearson Custom Publishing.

Mead, R. 1998. *International Management* (2nd ed). Blackwell: Oxford.

Corley, Robert. 2001. *The Legal & Regulatory Environment of Business.* (11th ed). Burr Ridge, IL: Irwin/McGraw-Hill.

Davis, D., Utts, J.M., & Simon, M. (2001) *Statistical and Research Methods for Managerial Decisions* [UOP Custom Edition]. 2nd ed. Ohio: Thomson Custom Publishing.

Lieberman, M. and Robert Hall. *Introduction to Economics.* Cincinnati, OH: South-Western College Publishing, 2001.

Kotler, P. (2000). *Marketing Management* (Millennium ed.). New Jersey: Prentice Hall.

Marshall, H., McManus, W., Viele, D. (2002). *Accounting: What the Numbers Mean.* McGraw-Hill: Student Study Resource, New York, New York

Nickerson, R., 2002. *Information Systems – A Management Perspective*. Pearson Customer Publishing.

Brealey et. el. (2002). *Fundamentals of Corporate Finance*. McGraw Hill, 3rd Edition.

Turban, E. et. al. (2002). *e-Business and Practices.* New Jersey: Prentice Hall.

www.ingramcontent.com/pod-product-compliance
Lightning Source LLC
Chambersburg PA
CBHW071423050326
40689CB00010B/1965